BSHS Monographs

1

L. J. Jordanova & Roy S. Porter, editors

Images of the Earth

Essays in the History of the Environmental Sciences

The British Society for the History of Science

First published 1979
Reprinted 1981

ISBN 0 906450 00 4

Published by the British Society for the History of Science, Halfpenny Furze Mill Lane, Chalfont St. Giles, Bucks., HP8 4NR, England

Printed by Alphaprint, Preston and the Central Printing Unit, University of Lancaster, England

CONTENTS

LIST OF ILLUSTRATIONS

Introduction

L. J. JORDANOVA & ROY PORTER

I

In the mid-1970s some historians of science working in Britain expressed disquiet with the development of their discipline. There was a feeling that research was becoming narrower and more specialised, and confined within routine and taken-for-granted intellectual boundaries. Partly as a response to this, some conferences were held (under the auspices of the British Society for the History of Science) to open up new ways of looking at the history of science, and to give a new direction to research. The first of these, held in March 1976, discussed "New perspectives in the history of the life sciences". Following it, a further conference was arranged for April 1977, at New Hall, Cambridge, on "New perspectives in the history of geology". It is out of the latter initiative that this book has emerged.

Our aim at the conference was to assemble scholars from many fields of research - history of science, history of religion, social history, history of art, practising geologists and cartographers, and museum curators, to discuss nearly thirty pre-circulated papers. The twelve papers in the present volume have all been developed out of papers originally given to the conference, though they have been revised for publication in the light of the discussions which took place at the conference and subsequently. We have chosen for this volume those papers which in our opinion formed the most coherent and challenging ensemble. We regret of course that some important areas are scarcely represented - such as the literary perception of the earth, and map making. But we hope that this collection will serve as a starting point, rather than an exhaustive record. The papers have been kept brief, being designed to pose new problems, challenge old assumptions, and sketch new lines of research, rather than to document incidents and achievements with copious displays of scholarship.

Our aim has been to move away from the biographical mode which has still been prominent in such recent collections of essays as Toward a history of geology (ed. C. J. Schneer, Cambridge, Mass., 1969). And, perhaps above all, we hope to stimulate thought about the earth broadly considered as an object of scientific investigation, transcending the conventional straightjacket of the 'history of geology'.

II

History of geology and of the earth sciences in general has only recently become a major area of specialisation among historians of science. Its origins, like those of so much history of science, are in the nineteenth century in the early histories of the field written by practitioners like Charles Lyell. (1) The aim of such works was to validate the intellectual debts of the self-consciously new science, to praise its founders, and justify its present form. Much history of geology has been done by those with training in the discipline, who frequently continue to practise their science with history as a subsidiary interest.

These two features, the connection of history writing with the early years of the science, and the continued involvement of geological practitioners, are important determinants of the shape of history of geology. Historiographical assumptions must also be seen in the broader context of changing attitudes to science and to the writing of its history. Perceived relationships between science and belief, between science and social utility, and between science and social roles, materially affect the history of geology we write.

Our purpose here is not to give an account of the history of the history of geology. Rather it is to indicate some major landmarks in the development of the subject in order to place the present volume in its context and to suggest avenues for further research. Considering the broad range of topics the book covers, the use of the term 'geology' is seriously misleading. We prefer to think of our subject matter as the history of systematic understanding of the earth. As a concept, 'the earth' is conveniently loose and flexible. It refers not only to the ground beneath our feet which geologists examine, but also to the environment, even to our planet as a whole. Furthermore it has played a major role in cosmological speculations in many cultures and, no doubt for closely connected reasons, it serves as one of the most powerful metaphors ever used. We would therefore expect to find historical evidence not just in the scientific treatise or field notebook, but in painting, map making, poetry, religious discourse, and so on. It is the achievement of several essays in this collection that they use such a wide range of primary materials. It is also clear that the history of the life sciences, of geography, and of the physical sciences, subjects closely related to the history of the earth sciences, must be mentioned to show the extent to which they have influenced work in the history of geology.

The scientific investment in history seen in the classic works of Geikie and von Zittel is evident, indeed it was characteristic of their generation. (2) Their clear conception of the limits and proper methods of geology determined their historical analysis; where they saw the coherence and unity of geology we would now see richness and diversity. In the sense that they wanted to know how people got things right, they were exponents of Whig history. (3) The assumed polarity between science and religious belief has presented a similar problem. This preconception must be seen in terms of the general nineteenth-century concern with faith and science which subsequently hardened into a scholarly tradition represented by Draper and White. (4) While there was thought to be a tension between science as such and religion, it was also believed that geology posed more of a threat to Christianity than, say, chemistry.

In the English-speaking world, historians have studied British and to a much lesser extent American sources. It should not be assumed that the relationship between geology and belief in Britain will be reproduced in Italy, France or Spain. Our vision has been determined by the British experience, but such assumptions should not prevent us from studying other countries in their own terms. To do this will involve turning to general history and other social sciences to understand different cultural and social contexts. To be precise, we could examine which features of science are culture specific and which are shared by different cultures at any particular time. (5)

We suggest that the very notion of geology has problems attached to it. It is, for example, by no means self-evident what geology includes, although it is arguable that anything called geology in the past should be included in the historian's purview. Similarly, the historian of the earth sciences would study any attempt to understand our planet without preconceived ideas about what the project should look like. These assumptions differ from the tenets of early workers in the field who carefully delimited the boundaries of the discipline. A consequence of the traditional approach has been that certain areas have been remarkably little explored from a historical perspective. Geography, physical anthropology, archaeology and palaeontology have all suffered neglect. Yet geology has frequently spilled over the tight guidelines drawn for it into these fields.

The situation may appear paradoxical. Geology was self-consciously defined as a new science in the late eighteenth and early nineteenth centuries, implying a restricted idea of what was 'properly geological'. At the same time, however, the discipline was bursting at the seams, leading investigators in other realms of nature to use a wide variety of approaches. Yet it is an important fact for the social history of science that distinct scientific disciplines did emerge, that in actual practice there were considerable and important overlaps between them while border skirmishes were also common. (6) We cannot explain such a complex phenomenon, but merely note that geologists increasingly felt that they had first option on the scientific study of the earth as an integral part of their professional self-image.

It has been established that the interests of geologists, and of those studying the earth before 'geology' existed, were far broader than a perusal of traditional history of geology would suggest. Nonetheless, the desire of both historians and geologists to draw such sharp boundaries must be explained.

Categorising and classifying the past is frequently an exercise which conceals more than it clarifies. There are several cases where the uncritical use of terms in the secondary literature must be subjected to rigorous analysis. The uniformitarian/catastrophist debate is the classic example in geology just as the vitalist/mechanist one is in biology. (7) It is clear that such a vocabulary acts in a constraining way by encouraging historians to see two clear-cut camps in opposition to each other, whereas in fact few of the individuals involved can be so neatly boxed. Probably the tendency to write history of science in terms of the contributions of great individuals has been even more damaging. It has gradually become more commonplace for historians of science to argue that by looking at the average man or woman of science one can gain a level of historical insight that cannot be found in the study of those we deem exceptional. The antithesis of the emphasis on a symbolic hero-figure is prosopography, which studies large numbers of individuals as groups in terms of social class, religious denomination, educational background, and so on. (8) This approach, particularly in its application to the environmental sciences, is still in its infancy. Nonetheless, because of its reliance on biographical data, it could be argued that prosopography is best supplemented by approaches which take account of the conceptual level of science. A growing interest in social groups is a healthy development which emphasises the concrete, human side of science and which may open the way for social analysis. It is, however, only one of the possible approaches to history of science.

To achieve a balance some attention should be paid to traditions in the history of ideas and intellectual history, which have provided scholarly models of great influence. What might be called the Johns Hopkins school of historians of ideas, initially George Boas and Arthur Lovejoy, and later Francis Haber and Bentley Glass as well, inspired those who sought to analyse concepts without looking simply at one figure. (9) And since ideas tend not to fall into convenient disciplinary boxes, they also encouraged people to see how views of the earth were intertwined with attitudes to mankind and to nature in general. Neither did they concentrate solely on British sources, but took full account of French and German intellectual traditions.

Taking such broad sweeps of ideas had inevitable limitations. There was a tendency to divorce ideas from their social context by roving across countries and periods, suggesting inherent continuity between ideas, and also by analysing the complex philosophical writings of an intellectual elite composed of important but possibly unrepresentative individuals. The reaction against this approach in the 1960s constituted what might be called a 'technicalising' of historical problems. Individuals, small groups, local institutions, and issues of geological technique were subjected to detailed investigation. The last ten years have seen an astounding increase of work in the field, but the social history of geology, and a full conceptual investigation of ideas of the earth, are relatively neglected. (10)

There is a new and very fertile tradition of history of ideas in the United States which is beginning to be taken up by European scholars. It has grown out of the intense interest that early Americans had in coming to terms with their environment, including the scientific study of it. A number of historians, going back to Henry Nash Smith in the 1950s have observed changing American attitudes to the earth and have tried to relate this to social,political and economic developments. (11) They have thus avoided the pitfalls of the Hopkins group, and at the same time they have greatly enlarged our understanding of natural history writing, of human tendencies to exploit nature, and of the role of the earth sciences in recent debates about conservation, to name but a few areas.

In part, the history of the earth sciences suffers from the limitations of history of science in general, itself a product of prevailing views of science which are increasingly under attack from sociologists, philosophers and historians. As the extensive literature on the subject testifies, history of science raises special historiographical problems, either not raised or already tackled and resolved in general history. (12) For example, it is frequently believed that one way to approach a social history of science is through its institutions. (13) Yet it remains common to document the history of one organisation in apparent isolation from society at large, something which is no longer acceptable in the broader historical community.

Some historiographical problems are undoubtedly more acute in the history of geology than in other areas of history of science. Despite the fact that geology was co-opted in the service of technology and industry, it has been assumed that geology was less socially involved than medicine, for instance. Its theories have been seen as neutral descriptions of objectively perceived data, prime examples being the use of maps and instruments. (14) There is growing dissatisfaction with such a naive philosophical position. Much attention has recently been paid, by psychologists, art historians and geographers among others, to cognitive mediations of reality emphasising the dissonance between image and reality. To take a classic case, maps do not neutrally represent the world but are drawn according to culturally determined conventions and are explicitly theory-laden. (15)

From a quite different tradition has come a view of scientific instruments as an integral part of theorising about nature, i.e. the instrument as an extension of the conceptual framework. This tradition, led by Gaston Bachelard, has largely studied the physical sciences where great attention has always been paid to the theoretical level. (16) Quite the opposite is the case with geology, which has been treated as a science concerned with the practical, tangible and the easily observable. This suggests the self-image of the science; manly, out-of-doors, communing with nature with an uncomplicated aesthetic appreciation of environmental beauty. This is obviously a stereotype although one perpetuated by many practitioners. It is important to remember the extent to which scientific disciplines actively promote such images and that each field has its own image. Physics, for example, has the image of an indoor, abstract, masculine science, while zoology and botany, although outdoor field sciences like geology, are considered distinctly more feminine. Scientific imagery and self-sustaining mythologies such as these are beginning to come under scrutiny and will, we suspect, lead to fruitful new approaches.

Geographers have used a similar approach in their analyses of how people, at different times and places, conceptualise their environment and their role in it. It is the subtle relationship between a human society and the landscape which surrounds it which has fascinated historians and philosophers of geography. Consequently, they have frequently studied the discovery of new environments, as in the exploration of Australasia or in the pushing back of the American frontier. They have used a variety of primary materials, poems, maps, paintings, philosophical treatises, novels, and political writings as well as the literature of science. Clarence Glacken might be said to be among the pioneers of the approach in his work which traced concepts of nature and culture from the Greeks to the end of the eighteenth century. His subsequent contributions have also helped to expand our ideas about how science helps human beings make sense of geographical variations. (17) More recently, Yi-fu Tuan and David Lowenthal have taken the approach further, using more literary evidence and giving more attention to the cognitive level. (18)

Glacken's work has shown how closely understanding of the earth was related to cosmological concerns. Specifically, he has drawn attention to the secularisation of attitudes to the environment and to the increasing assertiveness of the human right to mould and manipulate it. (19) Clearly this is a topic of great social and political concern, for instance in the rise of conservation and ecology movements in the last few years. Contemporary preoccupations have begun to inspire historical work, mostly on the policies towards the environment of the United States government. (20) This literature has not yet been fully assimilated by historians of science, although John Passmore's recent work in the area shows how fruitful the possibilities are. (21) There is already a sizeable literature on the related topic of geology and religious belief which highlights the social, moral and intellectual threat of the earth sciences to the established order. There is no doubt, however, that this extraordinarily complex field has been seen in terms of contemporary assumptions about geology and religion. It could be argued that to see the two as separate categories is to propagate serious misconceptions. Recent scholarship has unambiguously shown the intertwining of religious and cosmological theories, including theories of the earth, an insight much indebted to Hooykaas and further developed in the present volume. Despite its adherence to a historiography no longer current, Charles Gillispie's _Genesis and geology_ stimulated scholarly interest in natural theology and in the context of the Bridgewater Treatises in early nineteenth century Britain. The principal shortcoming of the work was its acceptance of inevitable tensions between _Genesis_, i.e. biblical literalism, and the emerging science of geology. (22)

The difficulty of defining clearly what may properly be denoted by the term geology at any particular time suggests that, as far as possible, historians should avoid using a vocabulary alien to the age under scrutiny. To take this to its logical conclusion is both difficult, and on occasion undesirable. The categories of our culture are so deeply ingrained and sometimes so necessary for analysis that an alternative mode of thought and the correlative vocabulary are almost impossible to find. Yet, to give an extreme example, to speak of 'geology' in the fifteenth century is to do such violence to historical reality that it should be deemed no longer acceptable. It is for this reason that we find the terms 'earth sciences' or 'environmental sciences' convenient alternatives since they are loose concepts which merely denote the objects of study.

A special case of these problems is the popular debate which has surrounded geology since the mid-nineteenth century where the issues have frequently been dramatised and distorted beyond recognition. Three areas are of paramount importance: the origin and age of the earth, the antiquity of man, and the theory of evolution. (23) While there is an extensive popular, even melodramatic, literature on the history of discoveries of early human remains, there have been few systematic attempts to understand the relationship of comparative anatomy, the sciences of mankind and palaeontology, to geological science. Nor is the origin of the human race the only place where these disciplines come together; they were intimately associated in the natural philosophy of the seventeenth and eighteenth centuries and in all evolutionary theories. The life sciences and the earth sciences are most closely bound together in the study of fossils, an area of history of science which is still seriously neglected. (24) Moreover, their common intellectual heritage is reflected in the social characteristics of the environmental sciences; practitioners in the life sciences, including the medically trained, were frequently keen geologists and meteorologists, while such shared concerns were further seen in institutions which taught and fostered research in both areas. We should note that geology was a popular science in another sense, in that it was pursued by vast numbers of amateur naturalists, especially during the nineteenth century. (25) The choice of geology for such activities at specific times and among certain social groups rather than botany or zoology remains to be explained. Ultimately this must lead to an analysis of social factors.

The social history of geology is beginning to take shape. Initially, it has involved the study of the process of professionalisation and of institutions. Both are aspects of past geology involving the external characteristics of organisational structure, social status, economic position, access to power, research facilities and so on. This approach tends not to touch the content of science which is so bound up with the social aspects, although in our view there is no reason why the two should not be studied together. Extending social history in this way certainly implies that social characters are linked to the 'scientific', i.e. that the kind of geology done is essential to understanding the institutions which house it, and _vice versa_. Furthermore both social and intellectual factors are necessary to understand the earth sciences in the wider society of which they are a part. There is resistance to this approach, especially from some members of the scientific community who hold that scientific knowledge, being value free, may not be analysed in social terms, although career structures, and so on, may. Here it is once again important to note the extent to which such beliefs affect historical preconceptions.

Thus, the history of the earth sciences, just like the history of science in general, has traditionally fallen into two classes. First, the concrete approach emphasises instruments, institutions, the objectivity of maps, and field-work. This has sometimes taken the form of listing advances in mapping or instrumentation techniques, or in writing accounts of a great institution or a great man, or of describing the practical expertise of past generations. More recently, the financial aspects of geological careers and their effects on day-to-day scientific life have been studied using models of professionalisation more indebted to sociology than to history of science. (26) Second, the intellectual approach looks at the writings of geologists as its primary datum, and has traced particular ideas or traditions as they 'progressed' over time or documented new discoveries and their role as a stimulus to advancing knowledge. Recent work has tended to take more circumscribed areas for study, using a shorter time period and a single scientific community. At its most philosophical such an approach examines systems of classification of rocks and minerals, using methods applied to other classificatory systems by Michel Foucault and others in French traditions of history of science who have been influenced by structuralism. (27)

But both social and intellectual methods in the history of the environmental sciences have avoided more fundamental political and economic questions which are commonplace in the general historical literature and indeed are increasingly so in other areas of history of science. The involvement of geology in the exploitation of the land for agricultural purposes or to supply natural resources for industry is a little touched field. Yet, the ideological role of science in early industry is of evident interest to scholars who want to analyse science and social structure. Doing this involves far more than combining the two prevailing approaches; it involves bringing to the specialised area of the earth sciences the tools developed by historians, indeed by social scientists generally.

We do not intend to suggest that geology is without its special features; on the contrary, historical work should attempt to reveal them. Our point is that the characteristics of geology, or indeed of the environmental sciences in general, cannot be taken as given or self-evident, they must be learned from historical materials. Neither should we have preconceptions about what could have influenced the development of the sciences. It seems reasonable not to rule out any factor as irrelevant to understanding past attitudes to the earth and the atmosphere.

The way historical problems are selected for research is obviously a crucial factor in determining historiographical patterns. We may even begin to feel that organising historical work by scientific discipline is unsatisfactory, so that rather than working on geology, a historian might choose to work on the professions, making geology one case among many, or on science teaching, in order to compare the ways in which the different sciences were integrated into the secondary school curriculum. This avoids the ahistorical method of drawing distinctions between scientific disciplines which in fact have no such boundaries. Indeed, several papers in the present volume have taken such a cross-disciplinary approach. The contributors focus on the eighteenth and nineteenth centuries in their attempts to expand the limits of the history of the earth sciences. Their work shows how ill defined those limits are and explores previously uncharted territories.

III

The division of the book into four sections represents a number of avenues which historians of the environmental sciences have begun to explore. The first section, 'geology and belief', offers reconsiderations on the theme which has been deeply controversial throughout the development of geology, that is, the relations of the science to belief. All three of the essays in this section deal with religious thought, specifically with Christian theology, but they mark a break with the traditional interpretations (of Genesis and geology, or geologists struggling with Moses) in not isolating religious from other forms of general philosophical belief about the nature of the universe and man's place within it. Ralph Grant's paper in particular explains how theories of the earth from Burnet to Lyell were articulated in terms of changing philosophies of mankind and his destiny.

Following on, above all, from the work of Hooykaas and Cannon, historians of geology have fortunately ceased to ridicule religion as a force hindering geological progress. (28) But they have hardly yet begun to do justice to the complexity and ambiguities of Christianity as a highly differentiated set of beliefs, changing markedly over time. Geoffrey Cantor's paper looks at the ideas about the earth of the High Church Hutchinsonian natural philosophers of mid-eighteenth century England. The Hutchinsonian philosophy is particularly challenging to modern historical categories, for it propounds a thorough-going mechanical universe, while also seeing nature as the emblem of the Trinity. In the same way, it holds up scripture as the ultimate authority in the interpretation of nature, while lauding a methodology of experience.

Similarly, John Brooke re-opens the issue of the relations between providence, miracles, and natural law in early Victorian England. These concepts were used to understand the earth considered as an object for natural theology. Brooke's paper demonstrates how natural theology cannot be taken as a monolithic discipline, and thereby places a question mark against attempts to see natural theology as the location of a 'common context' in the sciences of earth and life history in the nineteenth century.

The second section, 'The language and metaphor of nature', focusses attention on a gravely neglected field: how conceptions and theories about the earth are conveyed through languages, which themselves are carriers heavily laden (both overtly and unconsciously) with meaning taken from other disciplines. Following his earlier work in this area, (29) Martin Rudwick explores, particularly for the theory of Charles Lyell, the work done by metaphors and analogies borrowed from philology, Malthusian demography and political economy. He hints that conceptions of a natural history of language, which Lyell probably imbibed from German philology, constituted an important model for Lyell's notion of a statistical, 'populational' palaeontology.

Marcia Pointon deals with a different 'language' of geology: its visual language, as expressed by the school of mid-Victorian landscape artists who had been drilled into 'scientific painting' partly by the inspirational teaching of John Ruskin. Her essay underlines two powerful trends, which to some extent were working against each other. The first was the desire to create a scientific, geological art. Here she brings out the concern for precise depiction which animated landscape artists, not least the Pre-Raphaelites (but which led of course to works which differed radically from the scientific illustrations of geological books). But second, painters like Dyce were at the same time using geological formations as part of the living iconography of the human meaning of their art: the earth as a symbol of living nature, of antiquity, of the wearing hand of time, and so on. Here, as so often, human meaning and scientific meaning interpenetrate in ambivalent relations.

The third section, 'Geology and discipline boundaries', investigates why study and interpretation of the natural environment came to be broken down into distinct, specialist sciences. The base point is traced by Ludmilla Jordanova, who examines the attempt of scholars, especially physicians, in late eighteenth century France to construct a total science of the environment which was capable of explaining human disease and suffering. The aim of encompassing earth history and the ecological balance of flora, fauna and the human race, understood within a naturalistic framework of natural law, was an enterprise which characterised the late Enlightenment in Britain, Germany and the United States, as well as France.

That this bid had failed - as had Alexander von Humboldt's - by the middle of the nineteenth century is fully illustrated by the three other papers in this section. William Brock investigates the mid-Victorian interface between geology and chemistry via the polemical controversy between David Forbes and Sterry Hunt, in which (as in the debate with Lord Kelvin over the age of the earth (30)) geologists argued for the independence of the methods and laws of their own discipline against putative interference from a more 'basic', physical science. Patrick Boylan discusses disagreement over the early history of man amongst a scientific community split along two major axes: the divide between geologists and archaeologists, and, far more provoking, that between British and French men of science. David Allen then examines geology's relations in the latter part of the century with the field sciences. The starting point is the fact that geology was being overtaken in late Victorian England by newer, more attractive, sub-specialities. Geology was losing its previous hold on an amateur public, and geography was having far more success in making inroads into schools' curricula. By the early twentieth century the dream of a unified environmental science was further than ever from being realised.

The last section, 'The social history of geology', contains two pioneering essays in what has hitherto been perhaps the most neglected area of the history of the earth sciences. Through their detailed scholarship, both bring home the complexity of networks in the earth sciences, and the danger of making premature categorisations within inappropriate frameworks of interpretation. Hugh Torrens, in his meticulous study of geological activity in the Bath region in the second half of the eighteenth century, establishes what a wealth of interest existed in the broader aspects of the scientific study of the earth. But he also shows the inability - before the insight of William Smith - to convert this knowledge into guidelines of economic value to mining enterprises.

Paul Weindling also employs the theme of 'utility' to lay some queries against prevailing interpretations of the institutions most connected with early nineteenth-century British geology. He shows that we must avoid the temptation to presume that the role of the Royal Society in the development of British geology was 'reactionary' and that of the Geological Society of London was 'progressive'. The politics of the institutionalisation of geology, and of metropolitan/provincial relations were far too complex for such Whiggish terms to be useful probes, although the tensions between the two institutions were, in a very real sense, political.

Unlike John Woodward, the contributors to this volume would not claim that they have left no stone unturned. Much work, conceptual and empirical, is still to be done. For it remains a deplorable fact, how little historians have understood the myriad meanings for the human race of its own planet.

Notes

1 See C. Lyell, "Principles of geology, or the modern changes of the earth and its inhabitants considered as illustrative of geology", vol.i, London, 1830, of which the first five chapters constitute a historical survey. For analysis see R. Porter, 'Charles Lyell and the principles of the history of geology', Brit. J. Hist. Sci., 1976, ix, 91-103.

2 A. Geikie, "The founders of geology", London, 1897; reprinted New York, 1962; K. A. von Zittel, "History of geology and palaeontology to the end of the nineteenth century", trans. M. M. Ogilvie-Gordon, London, 1901; reprinted Codicote, Herts., 1962; H. B. Woodward, "History of geology", London, 1911.

3 H. Butterfield, "The Whig interpretation of history", Cambridge, 1931.

4 J. W. Draper, "History of the conflict between religion and science", London, 1875; A. D. White, "A history of the warfare of science with theology in Christendom", London, 1896.

5 T.Glick (ed.), "The comparative reception of Darwinism", Austin, 1974; M. Bloch, 'Toward a comparative history of European societies', in F. Lane (ed.), "Enterprise and secular change", Homewood, Ill., 1953, pp.494-521.

6 A. Lenane et al., "Perspectives in the study of scientific disciplines", The Hague, 1976.

7 R. Hooykaas, "Natural law and divine miracle: the principle of uniformity in geology, biology and theology", Leiden, 1963; M. J. S. Rudwick, 'Uniformity and progression: reflections on the structure of geological theory in the age of Lyell', in D. H. D. Roller (ed.), "Perspectives in the history of science and technology", Norman, Okl., pp.209-27; E. Benton, 'Vitalism in nineteenth-century thought: a hypothetic reassessment', Stud. Hist. Phil. Sci., 1974, v, 17-48.

8 S. Shapin & A. W. Thackray, 'Prosopography as a research tool in history of science: the British scientific community 1700-1900', Hist. Sci., 1974, xii, 1-28; L. Pyenson,'"Who the guys were": prosopography in the history of science', Hist. Sci., 1977, xv, 155-88.

9 A. O. Lovejoy, "The great chain of being", Cambridge, Mass., 1936; F. C. Haber, "The age of the world: Moses to Darwin", Baltimore, 1959; H. B. Glass, O. Temkin & W. L. Straus (eds.), "Forerunners of Darwin", Baltimore, 1959; G. Boas & A. O. Lovejoy, "Primitivism and related ideas in antiquity", Baltimore, 1935.

10 See for instance, C. J. Schneer (ed.), "Toward a history of geology", Cambridge, Mass., 1969.

11 H. N. Smith, "Virgin land: the American West in symbol and myth", Cambridge, Mass., 1950; L. Marx, "The machine in the garden", New York, 1964; R. Nash, "Wildness and the American mind", 1967;

revised edn., New Haven & London, 1973;
A. Kolodny, "The lay of the land", Chapel Hill, N. Carolina, 1975.

12 See R. M. Young, 'The historiographic and ideological contexts of the nineteenth-century debate on man's place in nature', in M. Teich & R. M. Young (eds.), "Changing perspectives in the history of science", London, 1973, pp.344-438.

13 For important and relevant examples see M. J. S. Rudwick, 'The foundation of the Geological Society of London', Brit. J. Hist. Sci., 1963, i, 325-55; J. Morrell, 'London institutions and Lyell's career, 1820-1841', Brit. J. Hist. Sci., 1976, ix, 132-46.

14 For an instance see A. K. Biswas, " History of hydrology", Amsterdam, 1972.

15 M. J. S. Rudwick, 'The emergence of a visual language for geological science, 1760-1840', Hist. Sci., 1976, xiv, 149-95; E. H. Gombrich, "Art and illusion", London, 1960.

16 G. Bachelard, "La formation de l'esprit scientifique", Paris, 1947; idem, "The psychoanalysis of fire", trans., London, 1964.

17 C. J. Glacken, "Traces on the Rhodian shore", Berkeley, 1967.

18 D. Lowenthal, 'Past time, present place: landscape and memory', Geog. Rev., 1975, lxv, 1-36; idem (ed.), "Geographies of the mind", Oxford, 1976; Yi-fu Tuan, "The hydrologic cycle and the wisdom of God", Toronto, 1968; idem, "Topophilia", Englewood Cliffs, N.J., 1974; idem, "Space and place: the perspective of experience", London, 1977.

19 C. J. Glacken, 'Environment and culture', in "Dictionary of the history of ideas", ed. P. P. Wiener, vol. ii, New York, 1973, pp. 127-34.

20 See note (11).

21 J. Passmore, "Man's responsibility for nature", London, 1974.

22 See Young, op.cit. (12); Hooykaas, op.cit. (7); Rudwick, op.cit. (7); J. H. Brooke, 'Natural theology and the plurality of worlds: observations on the Brewster-Whewell debate', Ann. Sci., 1977, xxxiv, 221-86; C. C. Gillispie "Genesis and geology", Cambridge, Mass., 1951; "Science and belief from Copernicus to Darwin", The Open University, Milton Keynes, 1974, esp. units 9-10, 'Natural theology in Britain from Boyle to Paley', and unit 11, 'Genesis and geology'; R. Rappaport, 'Geology and orthodoxy: the case of Noah's flood in eighteenth-century thought', Brit. J. Hist. Sci., 1978, xi, 1-18.

23 H. Wendt, "Before the deluge", trans., London, 1968. Cf. J. Bronowski, "The ascent of man", London, 1973.

24 Though see J. C. Greene, "The death of Adam", Ames, Iowa, 1959; M. J. S. Rudwick, "The meaning of fossils", London, 1972; P. Bowler, "Fossils and progress", New York, 1976.

25 D. E. Allen, "The naturalist in Britain", London, 1976; Harmondsworth , 1978.

26 See Morrell, op.cit. (13). D. H. Hall, "History of the earth

sciences during the scientific and industrial revolutions", Amsterdam, 1976, is a recent attempt to construct a social history of geology from a quasi-Marxist perspective. But it is based upon a narrow range of secondary materials and its judgement must be treated with caution.

27 M. Foucault, "The order of things", trans., London, 1970; W. R. Albury and D. R. Oldroyd 'From Renaissance mineral studies to historical geology, in the light of Michel Foucault's The order of things', Brit. J. Hist. Sci., 1977, x, 187-215; G. Gusdorf, "Les sciences de l'homme et la pensée occidentale", 6 vols., Paris, 1966-; F. Dagognet, "Le catalogue de la vie", Paris, 1970.

28 For Hooykaas, see op.cit. (7); W. F. Cannon, 'The Uniformitarian-Catastrophist debate', Isis, 1960, li, 38-55; idem, 'The problem of miracles in the 1830s', Victorian Stud., 1960, iv, 5-32; idem, 'The impact of Uniformitarianism', Proc. Amer. Phil. Soc., 1961, cv, 301-14; idem, 'Scientists and Broad Churchmen: an early Victorian intellectual network', J. Brit. Stud., 1964, iv, 65-88.

29 M. J. S. Rudwick, 'Poulett Scrope on the volcanoes of Auvergne', Brit. J. Hist. Sci., 1974, vii, 205-42.

30 J. D. Burchfield, "Lord Kelvin and the age of the earth", London, 1975.

Further reading suggestions

This brief list contains some titles, not in the history of geology narrowly conceived, but suggestive as regards the construction of images of the earth.

Blacker, C. & M. Loewe (eds.), "Ancient cosmologies", London, 1975.

Douglas, M., "Natural symbols: explorations in cosmology", London, 1970.

Duchet, M., "Anthropologie et histoire au siècle des lumières", Paris, 1971.

Eliade, M., "The myth of the eternal return, or cosmos and history", Princeton, 1954.

Eliade, M., "Myths, rites, symbols: A Mircea Eliade reader", ed. W. C. Beane & W. G. Doty, New York, 1975.

Foucault, M., "The order of things", trans., London, 1970.

Glacken, C. J., "Traces on the Rhodian shore", Berkeley, 1967.

Gusdorf, G., "La science de l'homme et la pensée occidentale", 6 vols., Paris, 1966-.

Humboldt, A. von, "Cosmos", trans. A. Pritchard, London, 1845-8.

Lowenthal, D. (ed.), "Geographies of the mind", Oxford, 1976.

Michelet, J. "La montagne", Paris, 1868, (on which see L. Orr, "Jules Michelet: nature, history and language", London, 1977).

Nash, R., "Wildness and the American mind", 1967; revised edn., New Haven & London, 1973.

Nicolson, M. H., "Mountain gloom and mountain glory", Ithaca, 1959.

Passmore, J., "Man's responsibility for nature", London, 1974.

Smith, B., "European vision and the South Pacific", Oxford, 1960.

Tuan, Yi-fu, "The hydrologic cycle and the wisdom of God", Toronto, 1968.

Tuan, Yi-fu, "Topophilia", Englewood Cliffs, N.J., 1974.

Tuan, Yi-fu, "Space and place: the perspective of experience", London, 1977.

Geology and belief

Revelation and the cyclical cosmos of John Hutchinson

G. N. CANTOR

It has become something of a commonplace to consider the eighteenth century as the 'age of empiricism'. It is widely believed that during this period British natural philosophers followed Newton's method of observation and experiment, and that they rejected other sources of knowledge. This attitude is supposed to have permeated all fields of thought, even theology, where men turned their backs on the Bible and employed empirical arguments to support either their natural theology or their atheism. While there may be some truth in this perspective, examination of the primary sources indicates not only that a considerable number of natural philosophers believed in revelation, but also, and more surprisingly, they consciously attempted to construct their science on the text of the Bible. Although some work has been done to elucidate the scriptural underpinnings of eighteenth-century science, historians have yet to ascertain the extent and depth of this relationship. In geology and cosmogony the connection is, perhaps, most manifest and it has already attracted some attention. Following Collier's pioneering study, some historians (such as David Kubrin) have examined the biblical cosmogonies of Thomas Burnet, William Whiston, and others. (1) Another example is the recent article by Michael Neve and Roy Porter which explores the relationship between scripture and field-work in the geology of Alexander Catcott (1725-79). (2) Yet Catcott was not unique in emphasising scripture as the major source of knowledge of the natural world. Many other writers throughout the eighteenth century adopted a similar approach. Whether these writers formed a 'movement' has yet to be ascertained, but many of those writing after about 1730 looked to John Hutchinson (1674-1737) (3) as their mentor, even if they were sometimes embarrassed by his polemics against Newton. There has been a tendency to ridicule Hutchinson's views; Leslie Stephen dismissed Hutchinson and his followers as 'whimsical writers, in whom a lawless fancy supplies the place of sound reasoning and enquiry'. (4) Nevertheless, Hutchinson was certainly an influential figure in British science and in geology in particular. His works deserve serious consideration (5) not only because of his wide influence, but also because the example of Hutchinson refutes many commonplaces about eighteenth-century science, such as the supposed dominance of empiricism and the rejection of revelation as a legitimate source of natural knowledge.

This paper attempts to analyse Hutchinson's writings, particularly his cosmogonical theory, in terms of his epistemological and linguistic theories. The paper comprises four sections. In the first, I discuss Hutchinson's theory of knowledge and attempt to define the roles he attributed to both empiricism and revelation. The second is a wide-ranging study of the differences between the 'Hutchinsonian' and 'Newtonian' world views, concentrating on Hutchinson's cosmogony. In the third, Hutchinson's views about language are analysed. Finally, some tentative connections are made between the language of the Bible and Hutchinson's theories in both science and theology.

I <u>Hutchinson's theory of knowledge</u>

The starting point for any discussion of Hutchinson's theory of knowledge is his attitude towards the Bible, and, in particular, the Old Testament. He considered that the Bible contained the word of God and that every word of scripture was perfect and contained the true description of nature's processes [C.xxix, H.28, L.90]. (6) However, he also believed that in the hands of man the scriptures had become debased and their true meaning obscured. For during biblical times the Jews had understood the Bible's full significance but subsequently had lost this knowledge which was intimately bound up with the original form of the Hebrew language. This corruption occurred at several related levels; they had lost their moral, social, and religious fidelity, their understanding of the physical universe and their knowledge of the unadulterated form of the Hebrew language. Instead, in order to assist the reading of the Bible they added 'points', i.e. dots, to distinguish different pronunciations of six consonants, and also the massoretic signs which introduced symbols for vowel sounds. With this degeneration of the written language, symbolised by such events as the 'Confusion of Tongues' [H.56], (7) and the loss of the physical theories (to be discussed in section II) inherent in the original text, men rejected revelation as the primary source of physical knowledge and instead framed fantastic hypotheses. According to Hutchinson, the only way in which complete knowledge of nature (and of morality) could be obtained was through the scriptures. However, he conceded that revelation was only one source of knowledge and that very limited aspects of the physical system could be ascertained by other means, for example, through the senses [H.85]. By contrast, he considered that the cardinal mistake of modern philosophers was their complete rejection of the method of revelation and their total commitment to other sources for their knowledge of nature.

In order to obtain the complete account of the natural world contained in the scriptures, Hutchinson had first to purge the Hebrew Bible of all errors of human origin, in particular, 'points' had to be eliminated. He had also to develop a sophisticated philological apparatus to interpret the Bible once it had been reduced to the original string of Hebrew consonants. In section III we shall discuss the way in which Hutchinson attributed meanings to these signs. However, one aspect of his semantics relates directly to this theory of knowledge, and can be approached through the following problem. There are many references in the scriptures to entities, such as cherubim, of which we have no immediate experience. How, then, can we understand this aspect of the Bible? Hutchinson claimed that we do not have intuitive knowledge of these entities [H.19, L.25]; instead, he appealed to what he called the method of comparison [C.xxiii]. Thus to form ideas of unobservable beings and their attributes we extrapolate from our ideas of material objects which we attain from our sensory experience. This mental process of comparison which played an important role in Hutchinson's theory of biblical exegesis emphasises the dependence of man's understanding of the scriptures on ideas derived through the senses. These dual aspects of his epistemology underscore his intention of reconciling science and theology in a constructive manner. Indeed, he considered that religions which fail to establish connections between revealed knowledge and natural knowledge would not last for long [C.xxxviii].

Sensory knowledge also played a more general role since Hutchinson subscribed to a broad-based empiricist psychology. He considered that 'our Senses were appointed Centinels to perceive and convey Ideas readily to us' of the material world, and ideas about matter and its motion in particular [P.4]. (8) In denying that the mind is supplied with innate ideas, which he linked with the philosophy of the atheist, Hutchinson considered that we attain essential aspects of our knowledge through experience by way of our senses. God had provided us with sensory apparatus in order to give us a further means of understanding the physical world and also to enable us to move around it with safety. Thus our sense of vision permits us to ascertain the position of bodies and their motions. While it may be inadequate to label Hutchinson simply as an empiricist, his theory of mind falls within a general empiricist tradition. Furthermore, he encouraged the empirical study of natural phenomena and he certainly made observations, if not experiments, in many branches of science, most particularly in geology. While Newton was 'living in a Box, peeping out at a

Window, or letting the Light in at a Hole', Hutchinson claims to have made numerous observations and to have collected thousands of fossils during his travels in order to illustrate the great processes described in the Bible [H.239-244]. (9)

The information from sensory knowledge was, however, circumscribed by three limitations. One of these was that the sense of vision gives us information only about macroscopic bodies, whereas the particles of the universal fluid, which is the cause of all motion (see section II), are beyond our perception. If we could see the particles of this fluid then our minds would be overloaded with ideas [P.8]. Secondly, the limited range of our sensory apparatus implies that for a full understanding of the hidden operations of the world - machine and for our spiritual and moral sustenance we must turn to revelation. Finally, our senses can deceive us; for example, when we are fooled by conjuring tricks [P.16]. In such instances our reason must prevail. This brings us to the third aspect of Hutchinson's theory of knowledge, the role of reason.

Throughout most of Hutchinson's writings the terms 'reasoning' and 'knowledge' were used in a pejorative vein to refer to the theory of knowledge adopted by free-thinkers. Such people, who committed the sin of pride, denied revelation and instead believed that 'they can discover all Divinity and Philosophy out of their own Heads' [L.115]. By proceeding in this fashion they were led into error and atheism and were thus destined to hell. In this context reason and revelation were considered by Hutchinson to be allegorised historically by the biblical account of eating at the tree of knowledge [H.47]. However, within his own theory of knowledge, reason, or what he usually called 'deduction', played an important role. Far from conflicting with revelation it was reason which made full study of the scriptures possible. The principal role of reason was to make inferences from the Bible - the fundamental 'data' of Christianity - about the physical world. Furthermore, reason not only allows us to study the natural world by comparing material things (e.g. fossils), but it also permits us with the aid of the Bible to make inferences from sensory data to the unobservable realm [H.20, L.18, 34-5]. Thus, in the method of comparison, deduction is the way in which we forge links between natural and revealed knowledge. Hutchinson acknowledged, however, that after the Fall deductions made by man can never be entirely free from error.

A further role for reason also involves the fourth and final aspect of Hutchinson's theory of knowledge, what he called 'relation', which may perhaps be translated as narration; that is, the writings of others. Reason allows us to assess the probability that any written account is true: 'we must set forth the Author's opportunity of knowing, abilities and means to know, interestedness, disinterestedness and capacity of relating what they saw or know' [P.156]. Hutchinson's own assessment of other writers ranged from those whom he cited in support of his own position to those who were positively incorrect or plainly ignorant. Among those favourably cited were many German, French and Dutch Protestant theologians who flourished in the first half of the seventeenth century, (10) while the villains of the piece on both theological and scientific grounds were Philo and his modern followers Isaac Newton and Samuel Clarke. Yet even the views of enlightened man were far inferior and more fallible than revelation. Thus free-thinkers were mistaken in pinning their faith on the writings and beliefs of other men. Indeed, one of the greatest dangers that Hutchinson saw facing both divinity and science was that the supporters of man-made systems of thought bent all evidence to suit their own trivial purposes [D.198, H.128].

In summary, Hutchinson's four sources of knowledge were (a) revelation, (b) relation (the narrative produced by men), (c) sensations, and (d) 'deduction'. The first category stands apart from the others on account of the importance and breadth of the knowledge it conveyed and also owing to its infallibility. Hutchinson considered these four sources of knowledge to be intimately related while he believed his opponents adopted one of the last three to the exclusion of the others and of revelation in particular. It must be stressed that for Hutchinson the choice of knowledge sources was a matter of prime importance. At stake was not merely man's knowledge of the physical universe but also the possibility of redemption. Anybody who failed to follow his own path to knowledge was, in Hutchinson's opinion, destined not merely to atheism but to hell.

II Hutchinson's cosmogony and his rejection of Newton's system

In discussion of Hutchinson's cosmogony, a functional distinction needs to be drawn between the processes involved in creation (as related in Genesis, and the recreation of the world during the Flood), and the post-creational state which involves the continuous cyclical action of a self-sustained world-machine. We shall concentrate principally on the latter set of processes. However, since this state represents the final condition of the creative process which Hutchinson discussed at length in *Moses's principia* [A], we must consider in outline Hutchinson's interpretation of the opening verses of Genesis. These verses, he considered, contained not only the true account of creation but also all the information, unobtainable in Hutchinson's view from other sources, which man requires in order to understand the formation of the world.

God created ex nihilo both the particles of inert gross matter, (which were to form the substance of earth, water and other macroscopic bodies) and also the far subtler atoms comprising the 'heavens'. These two substances were initially in a confused state and were probably located in different spheres, with the gross water particles tending towards the centre and the subtle fluid on the outside. Next, God endowed this subtle fluid in the form of 'spirit', which Hutchinson considered to be identical with air, with motion and thus activity. This moving air (or 'spirit') dried the earth and produced movement in its parts. Interestingly, Hutchinson alludes to the alchemical process of incubation as being analogous to the way in which the sphere of gross water particles was caused to solidify and differentiate. In the third verse of Genesis, Hutchinson encountered the action of 'light'. He argued that 'light' was a further modification of the subtle fluid which constituted 'spirit'; this time, however, the fluid was in a thin, expansive and moving state. This new activity compressed the gross matter into the shell of the earth which was both inscribed and circumscribed by uncondensed layers of water, while at both the centre of the earth and enclosing the system was the subtle fluid. God then caused the 'spirit' to expand, thus separating the passive gross matter into strata. This expansion also caused the earth's strata to undulate and crack; mineral veins were produced by this action and the waters were caused to circulate. At the extremity of the system the rarified subtle fluid came to form the firmament. Penultimately, God populated the world with all the necessary plants and animals, each containing its own kind of seed. Lastly, of course, came man.

The geological features of the earth were not discussed in much detail in Moses's principia [A, C, and D]. However, in a posthumously published work, A treatise on mining [S], Hutchinson dealt with specific geological formations, particularly strata, and such problems as the existence of fossils on mountain ranges. In this undated work he does not attempt to relate geological evidence to the text of Genesis, but considers that at the time of Noah's flood the power of gravity was suspended and the earth became dissolved into the waters. This catastrophic action by water and the subsequent deposition of strata largely accounts for the present state of the earth.

During the creative process, God acted as the prime mover but used the subtle atoms of 'spirit' and 'light' as intermediate causes which moved the particles of gross matter. After the creation, God no longer fulfilled this role but the operation of the world was maintained by the mechanical (i.e. contact) action of the universal subtle fluid (see section III). In place of God, the sun now acted as the source of all motion; (11) the fluid being moved by the sun and in turn moving the particles of gross matter. This fluid which composed the heavens was capable of three modifications: fire, light and spirit. The fire at the sun squeezed out particles of the fluid in the form of light which moved away from the sun towards the enscribing firmament. During this journey the motion of the particles gradually decreased and on reaching the firmament this motion was reduced to zero. There the particles congealed into larger groups, or what Hutchinson calls grains, which constitute the third modification of the fluid, that is, spirit or air. Once congealed into grains of spirit, the fluid travelled back from the firmament to the sun where it was broken down into its constituent particles by the fire and these were then once more projected toward the firmament. Thus a cyclical process occurs in which the quantity of the fluid is conserved. Furthermore, the process is self-sustaining, requiring no divine intervention. From this brief description of Hutchinson's cosmogonical theory we can see why he likened the operations of nature to that of a giant machine.

In order to appreciate Hutchinson's theories of matter and of power, an initial three-fold distinction will be helpful. Firstly, God exists (see section III for discussion of Hutchinson's use of the word 'Aleim' which for simplicity I shall translate as God), in whom infinite power resides. Secondly, there is the universal fluid constituted of invisible, inactive small particles which have been endowed by God with mechanical power and thus motion, which is conserved. Finally, there are particles of gross matter which likewise are inactive and not endowed with any intrinsic powers but instead are moved by impulses supplied by the subtle fluid. In section I,we saw that man's senses only give him information about gross matter and its movements while knowledge of God and of the world-machine composed of the subtle fluid can be obtained only from revelation.

The above distinctions show that real power resides only in God who imparted some power, but only of a mechanical variety, to the world-machine. This mechanical power is then transmitted by impulse between inactive particles of matter. According to Hutchinson, particles of both gross matter and of the universal fluid have only the properties of 'Solidity, Figure and Dimension [equivalent to extension?]' [H.94], but possess no means of affecting other particles except by contact action. In arguing for the importance of contact action and against action at a distance, Hutchinson pointed out that it is more to God's glory that he employs intermediate agents since mechanical action alone is comprehensible to man. Moreover, had God employed occult qualities, man could not distinguish between God's operations and those of the devil. This brings us to Hutchinson's central objection to Newton's natural philosophy. (Before turning to this, it should be noted that Hutchinson emphasised, even over-stated, the differences between his natural philosophy and that of Newton. By contrast, writers like Samuel Pike sought to reconcile 'Newtonian' science and the type of scriptural exegesis advocated by Hutchinson.) (12)

According to Newtonian dynamics, it is natural for a body once set in motion to continue moving. Thus Newton offered no efficient cause of either motion in general or planetary motion in particular. Furthermore, he associated with particles of matter certain active principles, such as gravitational attraction, by which one piece of matter affected another at a distance. (13) For Hutchinson, both inertial motion and action at a distance implied the existence of some power lodged in matter since neither form of motion was explained on the Newtonian system by contact action [H.96]. Yet, by lodging power in gross matter, Hutchinson believed that Newton had destroyed the fundamental distinction between God and the passive matter He had created. By endowing matter with power and thus by making it God-like, the Newtonians did not reconcile science and religion, but instead, according to Hutchinson, perverted each. Hutchinson complained that active powers in matter were mere fantasies not only because the Newtonians offered no evidence for their existence, but because the biblical description of nature left no room for such powers and instead attributed the motions of bodies to mechanism (see below). Furthermore, Hutchinson considered that any philosopher who attributed activity to matter would be led to 'Doubt the Veracity of the History of Creation and Formation and consequently' the existence of God [H.101].

In Hutchinson's cosmogony the immediate cause of all motion was the subtle fluid which filled all space. By contrast, the vast majority of Newton's universe was void space, the total quantity of matter being only enough to fill the metaphorical nut-shell. Yet void space conflicted with the scriptural account which stated that the universe was full [P.23]. By admitting void Newton had been forced to equate God with space, (14) or, as Hutchinson expressed the matter, Newton had 'patch[ed] up a God to constitute Space' [H.147]. This, however, was the God of the heathens and not the true God since the Bible stated that He is separate from the physical universe but He could both perceive and exert His power at any point in space [H.25, 184]. Moreover, by equating God with space Newton had once again impugned the basic distinction between God and the physical universe; he had limited God's power by requiring Him to act, and thus to be present, in every part of space [H.148-9, 184-5].

Hutchinson was well aware of Newton's ambivalent attitude towards subtle fluids and of his refusal to accept a plenum, which Hutchinson interpreted as an inevitable rejection of the mechanism described in the Bible. Newton's ether, which has sometimes been conflated with Hutchinson's universal fluid, was unacceptable on three counts. Firstly, it still permitted void space and action at a distance both of which were patently inimical to Hutchinson.

Secondly, the ether, unlike all other fluids, had the property of offering no resistance to moving bodies. However, argued Hutchinson, if it did not resist, then neither could it act as a mover to impel gross bodies [H.201]. Finally, and most importantly, by lodging power and activity in the ether itself Newton had in effect turned the ether into his God [H.136, 189]. Once again Hutchinson's objection turned on his need to maintain a sharp distinction between inactive matter and a powerful God.

Another issue on which Hutchinson's theories of the cosmos and of matter differed radically from Newton's concerned the conservation of matter and motion. Hutchinson considered that subsequent to the Creation the quantity of both matter and motion was conserved. His physical universe was circumscribed by the firmament where light particles congealed into spirit while at its centre the fire of the sun turned the grains of spirit back into light. In this closed system no matter or motion was either created or destroyed. The perfection of the world-machine in Hutchinson's theory contrasts dramatically with the Newtonian world view in which the quantity of motion was continually decreasing. This had important implications for both physics and theology; in particular, the specifically Newtonian doctrine of providence demanded God's immediate supervision of every part of the universe. (15)

What also impressed Hutchinson was the existence of numerous physical processes which offered close analogies to his cyclically-operating machine. Thus, for example, in both the circulation of the blood and the hydrological cycle the matter moved continuously in a closed circular path while the motion was maintained by a form of fire [H.23]. A further analogy, of some contemporary significance, which Hutchinson found most impressive, was the steam engine. Here gross matter underwent a cyclical process analogous to the one performed by the universal fluid in the world-machine. Thus fire in the boiler (analogous to sun) of the steam engine caused water (analogous to 'spirit', the universal fluid in the form of grains) to be rarified into steam (analogous to light, the moving particles of the universal fluid) and to circulate. At another part of the system, the condenser (analogous to the firmament), steam was condensed into water again [P.41, 71-85].

Hutchinson utilised the cyclical operation of the world-machine in order to explain a wide range of celestial and terrestrial phenomena. The following few examples will, I trust, suffice. The earth and other planets are caught between the outward flux of the light particles and the return flow of spirit towards the sun. These motions of the universal fluid cause the earth both to rotate on its own axis and to be carried round the sun [I.8, 31]. Likewise, the motion of all terrestrial objects, including plant and animal growth and even human locomotion, (16) was explained by the competing forces of expansion by light and of compression by spirit [P.47-68]. In this manner, although the specific details were often rather vague, Hutchinson offered explanations in mechanical terms of many of the phenomena which Newton attributed to gravity, attraction and active principles.

Hutchinson disagreed with Newton not only over specific scientific issues but more importantly he believed that Newton had approached the interrelation of science and theology in the wrong way. Instead of working down from the scriptures he had started with observations and had attempted to infer from them knowledge of a higher order. Yet, as discussed in section I, such a method could not lead to an understanding of how the world was created, the number of persons in the God-head, the immortality of the soul, etc [L.21-2]. Indeed, this method was bound to lead to the heresy of unitarianism. For Hutchinson, Newtonian natural theology was indistinguisable from the heathen worship of nature. Natural theology forced men to place too much emphasis on their own (fallible) reasoning and too little on God's words. Yet since neither Newton nor Clarke knew Hebrew, when they turned to the word of God they could only work from erroneous translations. Moreover, Hutchinson found the arguments of the natural theologians trivial and derisory; for example, he was bemused by Newton's argument (17) that the fact that we have two arms, two legs, etc, was clear indication of providential design [H.154, I.99]. Hutchinson believed that the universe was providentially designed but he denied that man could make inferences about God merely from observing His creation. Only once a man had gained salvation could he know the essence of God [H.63], but Newtonian natural philosophers, owing to their rejection of revelation, were destined to damnation, not salvation. (18)

III <u>Hutchinson's theory of language</u>

A recurrent concern among fundamentalists has been the analysis of biblical Hebrew in order to comprehend the hidden meaning of the scriptures. The cabbalistic schools of medieval Spain provide an early example. (19) Earlier this century, Benjamin Lee Whorf became interested in linguistics through a study of biblical Hebrew since he, too, believed that the true, but cryptic, meaning of the Bible could be obtained by attributing a specific inherent meaning to each letter of the Bible. (20) There were also many in the eighteenth century, including Hutchinson, who shared this preoccupation. A considerable proportion of his published writings are concerned with linguistics, and he developed a sophisticated, but ultimately untenable, theory of the Hebrew language which, it is suggested, played a major role in his world view. It would, in fact, be difficult to over-estimate Hutchinson's concern with language; even the very act of creation by God was a linguistic process. God made the world-machine 'not with a Hammer but with a Word' and fixed the parts together 'not with Iron, but by a Command' [A.16, D.175].

Unlike the natural theologians who used such metaphors as the divine artificer or clockmaker, Hutchinson's God was the divine linguist or penman. Thus, as recorded in Genesis, God gave different names to fire, light and spirit. The fact that He had named them differently implied to Hutchinson that they were not separate entities but merely modifications of the same fluid.

For Hutchinson the language of the Bible was perfect. The Hebrew words not only convey precisely the ideas of things, but there is also an exact correspondence between the words of the language and reality [A.16, C.xxix, H.28, L.90]. Furthermore, this correspondence involved two distinct levels of reality, the physical and the spiritual; for example, the same Hebrew word signified both the natural light, and the ineffable light [C.xxvii]. As discussed in section I, Hutchinson's empirically-based psychology implied that we can know only those objects which we have experienced. In order to decipher the Bible, we must infer the indirect significance of words from their material signification by means of the method of comparison [I.9]. Thus the scriptures are capable of two parallel interpretations each of which is true; for example, in Genesis, prior to the formation of man, one description refers to the physical creation, the other to God's activity [A.16]. Hutchinson considered that after the 'Confusion of Tongues' appreciation of this dual form of discourse was lost and so the heathens worshipped material objects themselves without realising that these objects were meant as signs signifying spiritual entities. Thus, the lion, one of the emblems representing Christ, came to be worshipped by the Egyptians [K.389]. In other words, the heathens in their worship overlooked the fundamental distinction, discussed in section II, between the spiritual and physical realms.

Central to Hutchinson's linguistic theory were two procedures which he used to establish relationships between the meanings of different Hebrew words. (21) One enabled him to assign a range of meanings to a specific Hebrew word, while the other connected different Hebrew words having the same 'unpointed' consonants. We shall examine in turn these two procedures.

In discussing any particular Hebrew word, Hutchinson usually cited the definitions given in Hebrew lexicons together with the biblical passages in which the word occurred. In this manner he was able to generate a wide range of associative meanings. Consider, for example, the sixth verse of Genesis which reads in the King James' version: 'And God said, Let there be a firmament in the midst of the waters . . .' Concerning the Hebrew word 'raki' (רקיע) here translated as firmament, Hutchinson refers to the concordances of Marius of Calasio (1621) and Kircher (1607) in which the word is taken to signify not only the firmament but also the expansive condition of some substance. The full range of this associative meaning is not denoted by any single word in the English language ('because we have no Idea of it'), but Hutchinson renders it as 'Expansion' [A.29-36, D.265-7]. It is this semantic connection between the firmament and expansion which underpins the functions attributed to the firmament in Hutchinson's cosmogony. Thus in his discussion of the creation, the 'spirit' was expanded to form the firmament. Likewise, in the subsequent operation of the world-machine 'spirit' (or air) has the property of expansion.

A further example is the word 'khoved' (כבוד) which according to some of the dictionaries Hutchinson consulted means 'to make heavy'. He extended this meaning to include the notion of to gravitate. The same word, in its unpointed form, also signified in some dictionaries the idea of glory, in the sense of glorifying a King [I.5-6]. For Hutchinson glory and gravity thus became related through language. Furthermore, they were also connected conceptually since light was the cause of gravitational attraction and light, in a metaphorical sense (or what Hutchinson called 'emblematically'), was the glory emanating from Christ. Similarly, in Christian art, glory was represented by light rays radiating from the head of Christ [I.26]. While Christ was linguistically and conceptually related to the universal fluid in the form of light, the other two modifications of the fluid represented the Father (fire) and the Holy Ghost (spirit or air). Thus the theological Trinity had its counterpart in the world-machine. Furthermore, Hutchinson pointed out that in the Bible several different words were used to signify, say, light. This, too, he considered significant since each of these different words indicated a distinct condition or mode of action of the light. Similarly the different terms used to refer to Christ indicated each of his different 'Offices' [C.xix, D.358, P.46]. Hutchinson considered that by solving linguistic problems of this type 'one Sense of each Word will run through the Whole, and the Science of Nature and Theology would strengthen each other reciprocally' [C.xxii].

The second procedure concerns Hutchinson's belief in the perfection of the Hebrew language. He considered that when words contain similar arrangements of 'unpointed' consonants, the things they signify must be related conceptually. The example I would like to discuss concerns some of the central concepts in Hutchinson's cosmogony. He analysed three similar Hebrew words:

שם	'sam'	he placed, put, disposed.
שם	'shem'	a name.
שמים	'shamaim'	the heaven(s).

In their 'unpointed' Hebrew form (first column) the words are similar and in the first two cases identical. Hutchinson considered these three words semantically interrelated if not conceptually congruent. Consider first the relationship he posited between 'sam' and 'shem'. He considered the first of these words to be a noun: the place or the space. Thus 'the Place and the Name are the same' from which he argued the proposition that 'Substance and Space are the same' [D.79, G.258]; a doctrine with a Cartesian ring about it.

Of even greater significance is the relationship he established between 'shem' and 'shamain' since this provided the crucial link between God's creative act and the subsequent function of the world-machine. The semantic connection is best illustrated in the following speculative passage in which he suggested that 'the Heavens were called by the Word used for [the] Names [of the Trinity] . . . and perhaps, שם which the translators have rendered Name, Gen. xi.4, might be an Image, or Representation of the Heavens, or of some Branch of Condition or Power in them' [D.102]. Thus in his own writings the two words were frequently interchanged; thus Hutchinson's translation of Psalm 19 reads: 'שמים the Names declare the Glory of God . . .' [F.207-8]. The implications of this semantic relationship can be seen with respect to Hutchinson's cosmogony as discussed in section II. We can now appreciate Hutchinson's rationale for considering the Trinity in the God-head to be analogous to the three conditions of matter in the heavens (i.e. fire, light and spirit), since both were signified by the same Hebrew characters. Furthermore, just as the spiritual Trinity has one essence, so, by analogy, the three forms of matter in the world-machine are of one substance [F.198].

The form of linguistic analysis employed by Hutchinson extended also to his method of deciphering emblems, hieroglyphics and even architecture. Thus, for example, Solomon's Temple and its ornaments, as related in the first book of Kings, were constructed on the same principles as the universe itself but on a microcosmic scale. For Hutchinson the Temple's structure represented the world-machine and the descriptions of each of its parts were emblems for the components and processes in the macrocosm [Q.1-86]. In adopting this approach, Hutchinson may be seen to be drawing on the extensive seventeenth-century concern with symbolism relating to alchemical emblems, Egyptian hieroglyphics, and the attempts to frame a natural language. (22) However, while Hutchinson may have adopted the techniques of the Cabbalists and Rosicrucians, my interpretation of his works distances him considerably from the magical tradition. Indeed, despite some residual elements of alchemy, he explicitly rejected a mystical approach to religion, in which, for example, miraculous power was vested in the Hebrew name for God [K.74].

For Hutchinson, the post-biblical dissociation of science and theology was intimately bound up with the loss of the original Hebrew language. Only in the semantics of the Hebrew language did words represent reality, while in all other languages words had only conventional significance. In the translation of the Bible into Greek, Latin and other imperfect languages, the rich layers of meaning were destroyed since these other languages differed from Hebrew both syntactically and semantically. Hutchinson appears to have had a strong sense of the cultural relativity of language [H.129]. Even individual Hebrew words, such as 'God' (see below), lost their original signification in translation and instead took on different meanings, which were associated with beliefs prevalent among the heathens who spoke imperfect languages. In science, too, 'senseless Words' were employed which failed to signify any of the real entities which were adequately described in the Bible. Indeed, Hutchinson complained that among contemporary philosophers 'nature' had become a 'Cant Word, without any Signification' [L.40, 144]. Furthermore, Hutchinson maintained that the current theories in science and theology were false because their proponents, and in particular Newton and Clarke, were dependent on erroneous translations of the Bible since they were unable to understand Hebrew.

One language which Hutchinson utterly rejected was mathematics. In this he opposed Newton, Descartes, and their followers who emphasised the role of mathematics in the analysis of nature. Hutchinson, who claims to have had a good grounding in mathematics, accepted that God had framed the world according to specific proportions. This is related in _Isaiah_, XL, 12, but other scriptural passages such as _Jeremiah_, XXXI, 37, (23) asserted that man must not measure the heavens and earth otherwise he would be disowned by God. Since the Bible is true, Hutchinson claimed that 'there can be no Application of _Mathematicks_' [H.226]. Yet his discussion of mathematics did not rest solely on biblical texts. Among his other objections to using mathematics to describe nature were the following:

(a) There were major disagreements among astronomers over celestial distances and magnitudes. Hence, far from being a precise language, mathematics was riddled with error and uncertainty.

(b) He rejected the view that mathematics was an important modern innovation and instead pointed out that the ancients had an excellent grasp of mathematics. This knowledge had been lost during the 'Confusion of Tongues' (_sic_) and was readily rediscovered in the seventeenth century.

(c) Hutchinson pointed out that mathematics was applicable equally to arguments with true or false premises. Newton (whose diagrams in the _Principia_ Hutchinson likened to 'Cobweb[s] of Lines and Circles to catch Flies in' [H.222]) had tried to construct his picture of the universe on mathematical principles. This, considered Hutchinson, was the wrong way of proceeding since he should have started with the biblical description of the world-machine and then attempted to set these in a mathematical form. Hutchinson suggested the type of explanation involved. According to his theory, the radiation from the sun decreases with distance; thus the motive power of the outer planets is less than the inner ones.

In this section, I have attempted to show that Hutchinson's analysis of the Bible was founded on a rational programme involving a sophisticated theory of language. Moreover, whatever the shortcomings of his linguistic and physical theories, the above discussion casts considerable doubt on the claim by Leslie Stephen and others that Hutchinson was merely a capricious crank.

IV The role of language in Hutchinson's cosmogonical theory

An interesting problem area, and one deserving further analysis, is the role which language plays in scientific theorising. Benjamin Lee Whorf has even suggested that we 'dissect nature along lines laid down by our native languages. This fact is very significant for modern science, for it means that no individual is free to describe nature with absolute impartiality but is constrained' by language. (24) While such speculations cannot be accepted at face value, we can, perhaps, attempt to answer a much more restricted question about the influence of language on scientific theory: was Hutchinson's cosmogonical theory affected by the language of the Bible? Before turning to this question we should note that Hutchinson did not propose his scientific theories in isolation from contemporary ideas. His discussion of the creation process did not differ radically from some of the other cosmogonies discussed by Collier, (25) and he was certainly familiar with the writings of John Woodward, William Whiston, and many others. Thus Hutchinson should certainly be viewed in this intellectual tradition from which he drew many of his concepts. However, while the role of language should not be over-stated, certain specific aspects of his scientific theories, and particularly his discussion of the post-creational state of the physical universe, indicate the influence of linguistic considerations.

It is reasonable to suppose that Hutchinson was socialised into the English tongue at an early age and then 'dialogued' with Hebrew, not in the form of a spoken language, but as a set of symbols, a form of code from which he deciphered a particular message. The method he used to decipher this code also affected his translation of the Bible. For example, as discussed in section III, Hutchinson turned to various dictionaries in order to generate the range of meaning of any particular Hebrew word. In emphasising this point, I wish to deny the suggestion that

Hutchinson's cosmogonical theories were simply 'read out' of the unpointed Hebrew text. To the contrary, Hutchinson's interpretation of the Bible was mediated through complex conceptual and philological structures.

The following considerations show the ways in which Hutchinson's cosmogony was shaped by the Bible.

(a) Hutchinson's investigations of the natural world were centred on deciphering the text of the Bible. Hutchinson was, of course, not unique in employing this source for knowledge about the creation of the world. Less typical is his insistence that the Bible should be used to discover the physical processes of the world in its post-creational state.

(b) In trying to explicate the message of the Bible, Hutchinson was aware that certain words could not adequately be translated into English. Thus at the lexical level certain 'hebraic' elements entered into the description of his physical theories. His use of unfamiliar theory-laden words such as 'the Names', 'Gravitor', etc, which makes his prose difficult to read, stems from this problem of translation. (This raises the yet unanswered question of whether Hutchinson's close study of Hebrew is reflected in his obscure style.)

(c) Grammatically, the connection appears to be quite strong. A recurrent example concerns words with the Hebrew ending 'im', which Hutchinson believed always signified plurals. Thus, he argued, since the word for water ('maim' - Genesis, I, 2) displayed the plural form, there must be two regions occupied by water, one inside the sphere of earth, the other beyond it [A.17]. Similarly, he related the plural form of the word for heaven ('shamaim' - Genesis, I, 1) to the three modifications of the universal fluid. Most significantly, he employed in his writings the word 'Aleim' (usually spelt 'Elohim') instead of 'God' since it too has a plural ending in Hebrew. He argued that those who ignored this grammatical point and instead translated 'Elohim' as 'God' or 'Deus' committed the heresy of attributing creation to a single being, whereas the scriptures were explicit about the Creator's plural - indeed tri-personal - nature. These three examples illustrate how Hutchinson's theories were affected by Hebrew grammar. Many other examples could be cited.

(d) Through the loose procedure of eliciting different English meanings of the same 'unpointed' Hebrew word, Hutchinson generated conceptual links within his physical theory. While Hutchinson chose only to emphasise certain of these connections, they existed in the text and in the linguistic tools that he used. Thus the key concepts of 'glory' and 'gravity' - and thus light as the cause of gravitation - were related through the word for firmament.

(e) His method of seeing a relationship between the meanings of words with similar Hebrew consonants also established specific concepts in his physical system. In particular, the example discussed above of 'sam' - 'shem' - 'shamaim' explains the relationship between several concepts central to Hutchinson's theory of the post-creational state of the universe.

In this paper, I have discussed Hutchinson's programme in natural philosophy in which he emphasised the Bible as the primary source of scientific knowledge. He did not entirely reject empiricism but attributed a specific and limited role to it. In order to decipher the message implicit in the biblical text, Hutchinson employed a number of ingenious methods of linguistic analysis. In this final and rather speculative section, I have suggested that while the language of the Bible was not the sole source of Hutchinson's science, it is justifiable to claim that linguistic considerations significantly fashioned his physical theories.

This paper does not take us far in solving the more general problem of the relationship between revelation and empiricism in eighteenth-century geology. In the case of Hutchinson, we see that revelation and empiricism were related in a complex and subtle fashion. Moreover, taken together with other recent studies, (26) we may perhaps appreciate more fully the diversity of approach available to the eighteenth-century geologist.

Notes

In preparing the final version of this paper, the author gratefully acknowledges the generous and helpful comments of Jonathan Hodge, Barbara Cantor, Chris Wilde, Alex Keller, Peg Jacob and Bernard Goldstein.

1 K. B. Collier, "Cosmogonies of our fathers: some theories of the seventeenth and eighteenth centuries", New York, 1934; D. Kubrin, 'Providence and the mechanical philosophy: the creation and dissolution of the world in Newtonian thought. A study of the relations of science and religion in seventeenth-century England', Cornell University PhD thesis, 1968.

2 M. Neve & R. Porter, 'Alexander Catcott: glory and geology', Brit. J. Hist. Sci., 1977, x, 37-60. See also R. Porter, "The making of geology: Earth science in Britain 1660-1815", Cambridge, 1977.

3 Among the better-known writers influenced by Hutchinson were Samuel Pike (1717?-73), William Jones (1726-1800), George Horne (1730-92), Duncan Forbes (1685-1747), and John Wesley (1703-91). See A. J. Kuhn, 'Glory or gravity: Hutchinson vs. Newton', J. Hist. Ideas, 1961, xxii, 302-22; R. E. Schofield, "Mechanism and materialism: British natural philosophy in an age of reason", Princeton, 1970, chapter VI; R. E. Schofield, 'John Wesley and science in 18th-century England', Isis, 1953, xliv, 331-40. The Hutchinsonian 'movement' is currently being studied by Chris Wilde of Darwin College, Cambridge.

4 L. Stephen, "English thought in the eighteenth century", 3rd edn., 2 vols., London, 1902, i, 389-92.

5 Hutchinson has been studied in those works cited above (notes 2 and 3) and also in H. Metzger, "Attraction universelle et religion naturelle chez quelques commentateurs Anglais de Newton", Paris, 1938, pp.197-200; A. W. Thackray, "Atoms and powers: an essay on Newtonian matter theory and the development of chemistry", Cambridge, Mass., 1970, p.246; P. M. Heimann, '"Nature is a perpetual worker": Newton's aether and eighteenth-century natural philosophy', Ambix, 1973, xx, 1-25.

6 Square brackets will be used to signify references to "The philosophical and theological works of the late truly learned John Hutchinson, Esq.", 12 vols., London, 1748-9. The code letter, which corresponds to the convention used in the index (volume 12), should be interpreted as below. Shortened titles are given followed by the volume number:

- A - "Moses's principia, part I", i.
- C - "Moses's principia, part II - Introduction", ii.
- D - "Moses's principia, part II", ii.
- F - "Moses's - Sine principio", iii.
- G - "A new account of the confusion of tongues", and "The names and attributes of the Trinity of the gentiles", iv.
- H - "A treatise of power essential and mechanical", v.
- I - "Glory or gravity essential and mechanical", vi.
- K - "The covenant of the cherubim: so the Hebrew writings perfect", vii.
- L - "The religion of satan, or antichrist, delineated", viii,
- P - "Glory or gravity. The second or mechanical part", xi.
- Q - "An inquiry . . . into the first temple of God built by Solomon", xi.
- S - "A treatise on mining", xii.

7 Hutchinson considered that the events at Babel were of religious significance and not related to the proliferation of languages

[G.11-12]. Cf. G. Steiner, "After Babel: aspects of language and translation", London, 1975, pp.57-62.

8 Hutchinson used the word 'Ideas' to refer to thoughts in general and he did not employ Locke's distinction between simple and complex ideas.

9 Cf. Neve & Porter, op.cit. (2)

10 For example, L. Capellus, S. Glassius, V. Schindler, S. Morin, S. Bochart, C. Kircher, G. J. Vossius and J. de Voisin. A useful list of Christian hebraists can be found in the "Encyclopaedia Judaica", Jerusalem, 1971, viii, pp.22-67.

11 M. C. Jacob suggests that Hutchinson may have been drawing on the alchemical tradition.

12 S. Pike, "Philosophia sacra: or, the principles of natural philosophy. Extracted from divine revelation", London, 1753, p.viii.

13 For discussions of theories of matter and motion held by Newton and his followers, see Schofield, "Mechanism and materialism", op.cit. (3); Thackray, op.cit. (5); Metzger, op.cit. (5); P. M. Heimann & J. E. McGuire, 'Newtonian forces and Lockean powers: concepts of matter in eighteenth-century thought', Hist. Stud. Phys. Sci., 1971, iii, 233-306. M. C. Jacob in "The Newtonians and the English revolution 1689-1720", Hassocks, Sussex, 1976, pp.61-5, links these active principles with the latitudinarians' commitment to social stability and their rejection of the mechanical philosophy which they associated with Hobbes and with social disorder. Hutchinson, although writing at a period slightly later than that discussed by Jacob, cuts across this distinction since he was committed to both social stability [H.6] and a mechanical model of the universe.

14 I. Newton, 'General scholium' to the "Principia", 2nd edn., London, 1713; Query 31 of his "Opticks", 4th edn., London, 1730. See also H. G. Alexander (ed.), "The Leibniz-Clarke correspondence", Manchester & New York, 1956.

15 D. Kubrin, 'Newton and the cyclical cosmos: providence and the mechanical philosophy', J. Hist. Ideas, 1967, xxviii, 325-46; P. M. Heimann, 'Conversion of forces and the conservation of energy', Centaurus, 1974, xviii, 147-61. The 'Hutchinsonians' were among those eighteenth-century natural philosophers who emphasised cyclical processes and conservation principles.

16 Hutchinson claimed that 'our Souls do not move the Parts of our Bodies; and the Brutes which have no Souls, move the Parts of their Bodies without them' [H.190]. This way he appears to circumvent the problems inherent in dualism, but he fails to appreciate the further problems which his solution raises.

17 Newton, "Opticks", (based on the London, 1730, 4th edn.), New York, 1952, pp.402-3.

18 Despite Hutchinson's extensive disagreements with Newton, recent studies of Newton's unpublished writings have exposed a number

of areas of agreement. See for example, Kubrin, op.cit (15); B. J. T. Dobbs',"The foundations of Newton's alchemy or 'The hunting of the greene lyon'", Cambridge, 1975, chapter V; J. E. McGuire & P. M. Rattansi, 'Newton and the "Pipes of Pan"', Notes Roy. Soc. Lond., 1966, xxi, 108-43. Newton and Hutchinson were also both interested in the language of the Bible and in assessing the significance of Solomon's Temple.

19 G. Scholem, "Major trends in Jewish mysticism", London, 1955, and his article 'Kabbalah' in the "Encyclopaedia Judaica", Jerusalem, 1971, x, pp.490-654. For eighteenth-century attitudes towards the Bible see H. W. Frei, "The eclipse of Biblical narrative. A study in eighteenth and nineteenth-century hermeneutics", New Haven & London, 1974.

20 J. B. Carroll (ed.). "Language, thought and reality: selected writings of Benjamin Lee Whorf", Cambridge, Mass., 1956, pp.7-9. See also P. C. Rollin, 'The Whorf hypothesis as a critique of Western science and technology", Amer. Quart., 1972, xxiv, 563-83.

21 Neither form of argument is generally valid. However, Hutchinson probably hit on some linguistic relations which are today considered significant.

22 See, for example, Scholem, "Major trends", op.cit. (19); M. David, "Le débat sur les écritures et l'hiéroglyphe en XVIIe et XVIIIe siècles et l'application de la notion de déchiffrement aux écritures mortes", Paris, 1965; J. Knowlson, "Universal language schemes in England and France, 1600-1800", Toronto & Buffalo, 1975.

23 "Isaiah", XL, 12: 'Who hath measured the waters in the hollow of his hand, and meted out heaven with the span . . . '; "Jeremiah", XXXI, 37: 'If heaven above can be measured, and the foundations of the earth searched out beneath, I will also cast off all the seed of Israel . . . '

24 Carroll, op.cit. (20), pp.213-14. For discussions of the Whorfian hypothesis see M. Black, "Models and metaphors", Ithaca, 1962, pp.244-57; I. D. Currie, 'The Sapir-Whorf hypothesis', in J. E. Curtis & J. W. Petras (eds.), "The sociology of knowledge: a reader", London, 1970, pp.403-21.

25 Collier, op.cit. (1)

26 Porter, op.cit. (2), and R. Rappaport, 'Geology and orthodoxy: the case of Noah's flood in eighteenth-century thought', Brit. J. Hist. Sci., 1978, xi, 1-18.

Hutton's theory of the earth

R. GRANT

This paper is principally concerned to offer a reading of James Hutton's _Theory of the earth_.(1) In the first place, it discusses that aspect of the text, the teleological, which constitutes the theory of the earth as theory in Hutton's own understanding of the term. Secondly, Hutton's _Theory_ is treated as part of the tradition of cosmogony, as Hutton explicitly acknowledged in chapter three of that work. In the third place, it treats the _Theory_ as an integral part of Hutton's writings considered as a unity.

A grasp of the meaning of the _Theory_ presupposes a grasp of a central theological problem which, although barely visible in the text, is nevertheless constitutive of the theory. The identification of this problem is at the same time the identification of the rationale for the pervasive teleological language so characteristic of Hutton's particular form of written expression. It is the need for an adequate appreciation of this central problem which demands reference to Hutton's other writings. (2) For although there are perspectives from within which the _Theory_ may be treated autonomously, these are in fact bound to result in the 'theory' proper remaining untouched. In so far as one is to deal with the theoretical language of the text, the textual divisions within the Huttonian corpus are merely contingent and, if considered to be more than this, become barriers cutting off the natural flow of the language and resulting in a reading which is semantically undernourished.

The truth of this claim is evident from the Hutton literature of recent years. A striking example is the demonstrated continuity, at the level of natural philosophy, between Hutton's chemical and physical theories and the dynamics of his geological system. Again. from the standpoint of social history, it has been Hutton's system of thought as a whole which has served as the medium for Roy Porter's identification of Hutton as a geologist whose speculative interests make him typical of the cosmogonical tradition - as relocated in the distinctive milieu of the Scottish Enlightenment - but less typical of the body of men who contributed to the growth of the geological sciences in Britain. (3)

The present paper develops this contextual approach. And although a recognition of Hutton's theological interests informs the studies which have build up this approach, it is this side of his thinking which still awaits a more systematic presentation. For it is precisely the attempt to deal with Hutton's theology and its relation to the theory of the earth that seems most prone to question-begging, as far as interpretation is concerned. The essential point is that Hutton's texts must themselves define his theological priorities, and thus display the manner in which these are served by his theorising in other fields such as the study of the earth. In particular, they must be allowed to define the specific theological problems they are concerned with. For it is possible to make Hutton's theology the centre of interest, emphasise his teleology in order to document the nature of his theological commitments, and yet, owing to the empty generality of the question informing one's initial viewpoint, inevitably leave off the interpretation before the meaning has been fully drawn off from the teleological stratum of the text. I shall now distinguish the possible levels of interpretation of a work such as the _Theory_, and then continue with an analysis of the theory itself.

We may call the first level the 'scientific', bearing in mind here Hutton's own distinction between 'science', as the formulation of 'laws' of nature of relatively restricted scope, concerned with the explanation of particular classes of natural phenomena, and 'philosophy', as the formulation of 'theories' which unify scientific laws into one 'system of nature'. Hutton's _Theory_ is a 'philosophical' work but with an identifiable 'scientific' content. The _Theory_ is therefore directly accessible to interpretation grounded in an understanding of such disciplines as stratigraphy and mineralogy as we know them today. The _Theory_ is accessible, in conjunction with such a work as the _Dissertations on different subjects in natural philosophy_, to an interpretation grounded in an appreciation of the problems of the geophysical and geochemical processes which ultimately determine the shape of geological phenomena. But already this is approaching a second level of interpretation because of the complexity of the general philosophy of nature which lies at the base of Hutton's system of physics.

Questions of epistemology, methodology, metaphysics, and theology must all be raised, and it is at this point that the _Investigation of the principles of knowledge_ appears as the key text. (4) All these questions have a bearing on the 'philosophical' level of the _Theory_.

Now it is clear that a 'scientific' reading is of no service at this level of interpretation. While valid in itself, it can have nothing to say on the matter of Hutton's teleology because it has not the prior interest in the questions which the teleological language expresses, and lacks the horizon within which to find this form of expression meaningful, seeing it only as the outpouring of Hutton's speculative excesses. But neither does a more positive attitude towards Hutton's speculative concerns lead far enough if the teleology is regarded as a mere surface layering on the geological content proper of the _Theory_, as is implied in much of what is said concerning Hutton's 'deism'. (5)

Hutton's teleology is not exhausted in the simple anthropomorphism of the world being 'designed' to benefit mankind, nor in the simple notion that the 'order' of nature reflects the 'wisdom' of the Creator. Rather, we must examine the relation of Hutton's teleology to the idea of the soul. In Hutton's system of thought a relation of analogy serves to distance the idea of the soul from the idea of nature. Furthermore, it is the soul which is the basic analogue in the relation, the 'active principles' or 'powers' of the natural order which are the objects of study in natural philosophy being merely its analogates. And Hutton put the analogy to use in a specific way, making the idea of nature serve his arguments for the immortality of the soul. This is the context in which the teleological language of the _Theory_ must be read if its final meaning is to be grasped. I shall now examine this language in more detail.

A theory of the earth, Hutton stated, 'should bring the operations of the world into the regularity of ends and means' (6) and, specifically, should demonstrate the means by which the earth is adapted to the purpose of being a habitable world. It is such an adaptation of means to ends which Hutton calls 'wisdom', the object of a philosophical 'theory'. In detail, Hutton was concerned to show that the 'mineral system', the subterranean realm of 'mineral fire', serves to maintain a constant area of habitable land through the consolidation and uplift of the sediment derived from eroding land-masses. It serves to maintain, that is, the 'terrestrial system' on which depends the 'vegetable system' on which in turn depends the 'animal system':

> It is therefore a proper view of the necessary connection and mutual dependence of all those different systems of changing things that forms the theory of this earth as a world, or as that active part of nature which the philosophy of this earth has to explore. (7)

It is this harmony between the various sub-systems of the system of the world at large which Hutton described as a case of 'means' (i.e. geological events) wisely adapted to a benevolent 'end'(the provision for plant, animal and, ultimately, human life). But it is just as important to note here what it is that he was concerned to deny as it is to note what he would affirm. His theory was intended to show a 'system of nature, in which may be perceived no ineffectual operation, nor any destructive intention, but the wise and benevolent purpose of preserving the present order of this world'. (8)

In the first place, let us consider the notion of 'destructive intention', an important example of which is vulcanism. Hutton was concerned to deny that volcanoes, because of their destructive effects, symbolise the presence of evil in the natural order. He used a traditional style of argument to the effect that a volcano can be seen as a manifestation of natural evil only if its action is seen as an end itself, whereas evil is rendered merely 'apparent' if volcanic action is seen as a wisely realised but subordinate end, and thus ultimately a means toward the realisation of a higher purpose. The end of volcanic action is the release of excess heat from within the earth, while the higher end of the operation of this heat itself is of course the 'habitability' of the earth:

> The end of nature in placing an internal fire or power of heat, and a force of irresistible expansion, in the body of this earth, is to consolidate the sediment collected at the bottom of the sea, and to form thereof a mass of permanent land above the level of the ocean, for the purpose of maintaining plants and animals. (9)

Natural philosophy in general is concerned with ends and means in nature. But another way of expressing this is to say that natural philosophy is concerned with the 'powers' or 'active principles' which make up the 'active part of nature', (10) and in this sense the wisdom of the natural order is to be discerned 'in the proper adapting of powers to an intention'. (11) In the theory of the earth in particular, the power of nature which must be seen to be properly adapted to its end is the 'internal fire or power of heat' referred to above. Its action, Hutton claimed at the outset of the _Theory_, has not been understood 'whether with regard to its efficient or final cause'. (12) And it is a consideration of this internal power of heat, _qua_ efficient cause or means, which arises when we turn to the second denial which Hutton made in the passage quoted above

- his denial of any 'ineffectual operation' in the system of nature. (13) That is to say, there is no power of nature whose action tends to no purpose or is insufficient for the effective realisation of the purpose to which it is directed.

Having examined the teleological structure of Hutton's theory, we can return now to the question, raised previously, of the relation of the theory to the idea of immortality. It is in terms of this idea that Hutton's system of thought is truly unified, and in the _Investigation of the principles of knowledge_ he not only dealt with the idea itself in great detail, but made it clear that it is principally this idea which is to be served by the unification of the various branches of natural and moral philosophy.

Hutton's idea of immortality depends on the further idea of progress. The term 'progress' in Hutton may refer either to the idea of social progress, the progress of the individual nation toward civilisation, or to the idea of the progress of the individual soul toward a state of moral perfection. The two ideas are closely related, however, in the sense that the perfection of the individual can be realised only through membership of a moral community of a special type, the emergence of which is itself one with that of the civilised state. But Hutton distinguished between two levels of perfection: the finite teleological perfection which marks the realisation of man's natural end as a morally responsible being, acting in society according to the dictates of his own conscience, and the 'absolute' or 'metaphysical' perfection of God. His idea of immortality depends on this distinction in the following way: if the natural end of the strivings within the human soul can be believed to be ultimately the transcendent ideal of perfection attributed to the divine nature, and if it can be believed that there is a certain recognisable 'mark' of this upon the soul in its present state such that it does appear to be in the course of realising this further end, then it may be believed that the soul is immortal. The soul is in the course of realising an end which it can never reach, and can only progress 'for ever' towards.

Hutton was here building upon a notion common enough in the eighteenth century, the notion that the human soul is possessed of capacities which cannot possibly be fully realised in the course of a mortal life-span, and that the soul is capable of 'infinite' improvement. For Hutton, the mark of immortality upon man's progressive nature was in the end aesthetic: progress in rational understanding and progress in moral awareness are all means of satisfaction of the primary appetite within the soul, the appetite for the forms of beauty in all its manifestations - natural beauty, moral beauty, and divine beauty; and it was Hutton's conviction, furthermore, that this is a form of satisfaction peculiar to the soul in itself:

> If there is such a thing as an enjoyment purely spiritual, in having no immediate dependence on sense and perception, and purely intellectual, as having in itself the principles of its rational gratification, it is here that we may fix the beginning of a mind which is not animal, and which, of consequence, is not mortal. (14)

And the point is that it is just a beginning: in life this capacity for enjoyment in the pursuit of knowledge 'is never perfect; for, the utmost perfection of man is to know his ignorance, in discovering an indefinite field for knowledge to which he has not yet attained'. (15) Hutton went on to raise the question of whether man can attain to this indefinite compass:

> Endless is this book of knowledge which man has been made to understand; infinite is this field of refined enjoyment which he has been made to enter; and must he then be made only to know the character, without being suffered to read the work - only to see the means of fulfilling his desire, without being suffered to quench his thirst, in drinking at this source of intellectual delight? Must he just taste this fruit, delicious in itself and made for his enjoyment, only to regret his loss - only to know that his desire shall not be satisfied? (16)

Needless to say, Hutton denied this possibility. But what could he do to underscore his idea of immortality in rational terms in order to provide more than a reiteration of his own sublime enthusiasm which was after all the source of his belief? We must note that Hutton nowhere adopted the pretence that the immortality of the soul can be 'demonstrated'. Not only did he reject the ideas of bodily resurrection and of transmigration of the soul, but also any such ideas of the soul's natural immortality as depend upon the putative demonstration that the soul is possessed of an immutable essence. But he did consider that the immortality of the soul could be made the object of what he called a 'reasonable conjecture', and it is here that his teleology of nature came to play its part.

Hutton cast the idea of the soul into teleological form, equating the concept of progress with the teleological concept of realisation. On this basis he could attempt to reinforce his argument for immortality by turning to his systematic appraisal of purposiveness in nature, and thus to his claim that all active agencies or powers do fulfil the ends of their actions, and argue, by way of analogy, that the same is true of the powers of the mind. There is a fundamental difference between the powers of nature and the powers of mind: the actions of the former are bounded by finite ends which can be seen to be realised, while the field of action of the latter is open-ended and, in the case of the higher intellectual and aesthetic powers, infinitely so. But the relation of analogy still holds, and not only because Hutton in fact deliberately shaped his model for natural powers by analogy with his teleological psychological model. His basic belief was that the two realms are connected through their ultimate grounding in the power of God, and the point of the analogy is that a demonstration of the divine wisdom in the exercise of the powers of nature will, assuming the consistency of the divine nature, provide assurance of its control in the mental realm as well. If the powers of nature are seen to realise the ends of their actions without exception then, by analogy, the end or purpose of the human soul can be supposed to be realised also. That there are no 'ineffectual operations' in nature is thus a crucial point to be established, and Hutton's teleology here found its rationale in the attempt to eliminate systematically from the realm of nature any precedent for a failure in finality.

All Hutton's natural philosophy is tied up in this programme, but the theory of the earth occupies a special place. As Hutton wrote of the need to see the finality inherent in the action of the earth's 'subterranean fire':

> This subject is important to the human race, to the possessor of this world, to the intelligent being Man, who forsees events to come, and who, in contemplating his future interest, is led to inquire concerning causes, in order to judge of events which otherwise he could not know. (17)

This preoccupation with the future leads, as is well known, to Hutton's principle of uniformity.

> It is only from the examination of the present state of things that judgements may be formed, in just reasoning, concerning what had been transacted in a former period of time; and it is only by seeing what had been the regular course of things, that any knowledge can be formed of what is afterwards to happen. (18)

With respect to the earth, Hutton's belief was, of course, that the present order of things would be maintained and, as he concluded the *Theory* in its first published form, 'we have got enough; we have the satisfaction to find, that in nature there is wisdom, system, and consistency' and, accordingly, 'no prospect of an end' to its operations. (19) Hutton's theory of the earth thus provided him with an image of the unceasing action of the powers of nature maintaining the known form of the earth through an indefinite future duration, an image which has further significance:

> It is thus that enlightened natural history affords to philosophy principles, from whence the most important conclusions may be drawn. It is thus that a system may be

perceived in that which, to common observation, seems to be nothing but the disorderly accident of things; a system in which wisdom and benevolence conduct the endless order of a changing world. What a comfort to man, for whom that system was contrived, as the only living being on this earth who can perceive it; what a comfort, I say, to think that the Author of our existence has given such evident marks of his good-will towards man, in this progressive state of his understanding! What greater security can be desired for the continuance of our intellectual existence, - an existence which rises infinitely above that of the mere animal, conducted by reason for the purposes of life alone. (20)

Having looked at the way in which Hutton's idea of immortality informs, through the medium of teleology, his theory of the earth, we may turn to look at the historical relations of his theory at this eschatological level, where Hutton himself located the primary debate between himself and other theorists.

In chapter three of the Theory of 1795, 'Of physical systems, and geological theories, in general', Hutton, having insinuated that there had been no genuine theory of the earth in the past, went on to make it clear that his objections to past theories were aimed principally at the 'philosophical' or teleological level and not the 'scientific' level. Taking a narrow view, we may see that the theories have explanatory power but 'so far as the theory of the earth shall be considered as the philosophy or physical knowledge of this world, that is to say, a general view of the means by which the end or purpose is attained', they are radically deficient. (21) Clearly this is a loaded criticism, containing a demand that Hutton's own theological priorities should provide the criteria of assessment. Nevertheless, we can recognise Hutton's genuine engagement with the concerns expressed by these theories, as in the case of Thomas Burnet to which I shall now turn.

Thomas Burnet wrote his Sacred theory of the earth (1681-89) to serve a millenarian eschatology according to which the histories of man and of the earth are integrated in a chronological succession of discrete events: the creation of the earth, the paradisical state of nature or golden age of man, the destruction of this state by the deluge, the post-diluvial and present day course of nature, the coming conflagration, first resurrection (of the saints), earthly millennium, second or general resurrection and last judgement, and consummation of the earth. Burnet's theory of the earth was intended to account for the apocalyptic happenings supposed to befall the earth through both causal explanation and providential interpretation.

In the first published part of his Sacred theory (1681; 1684), Burnet dealt with the creation or 'original' of the earth and the deluge, making it clear however that this was but part of a more general theory which would 'present . . . under one view . . . the several faces of Nature, from First to Last, throughout all the Circle of Successions'; it would 'reach to the last period of the Earth and the End of all Things'. Burnet's conception of the future was grounded in his interpretation of prophecies found in the scriptures, and his 'causal' account of apocalyptic events is only partly intelligible in philosophical terms. These events are to result from an extraordinary concatenation of the natural agencies which determine the ordinary course of nature, particularly the agency of 'fire'. It was to this 'unsystematic' treatment of natural agencies that Hutton objected, particularly the construal of the agency of 'subterranean fire' as essentially destructive. For Hutton, this fire had to be interpreted as part of the mechanism maintaining the ordinary course of nature, the notion of extraordinary providence dismissed, and his own philosophy of fire caste into teleological form to provide the basis of an alternative providentialism.

The contrast between Hutton and Burnet can be highlighted if we consider for a moment their respective uses of the idea of 'succession'. In Burnet's theory the history of the earth was made up from a succession of events megascopic in nature: alternating catastrophic episodes and intervening stadia. The notion of 'revolution' was incorporated into that of succession as an irruption into an otherwise stable order. In Hutton, on the other hand, the idea of an 'order of succession' had cyclical overtones, absorbing the idea of revolution which meant only the turn over of a bounded state of affairs. These two opposing schemes may however be combined when the former is repeated indefinitely, giving an overall cyclical pattern of creation, preservation, destruction, creation, etc. This contrasts with the former scheme in being 'eternalistic' (taking the word 'eternal' in its loose sense), while it contrasts with the Huttonian idea of succession according to which the 'creative' and 'destructive' elements are held in tension as the ground of a balanced or 'steady' state. Lyell found this a source of some ambiguity when writing his history of cosmogony, saying that in the Indian, Egyptian and Greek sources the idea of a destruction of the world 'sometimes . . . would seem to imply the annihilation of our planetary system, and at others a mere revolution of the surface of the earth'. For example, in reference to the three persons of the Hindu deity, the powers of creation, preservation, and destruction, he wrote: 'the co-existence of these three attributes, all in simultaneous operation might well accord with the notion of perpetual but partial alterations finally bringing about a complete change'. (22) Such a steady state model was used in different forms by both Hutton and Lyell, although the latter thought that Hutton subscribed to the model of 'alternating periods of disturbance and repose' rather than continual balance. (23)

Hutton dispensed with the idea of creation in the sense of a beginning in time, preferring the idea of preservation or 'continual creation' of the earth through an 'endless succession' of cycles of decay and renewal. And there was a similar detachment from the traditional Christian idea of 'origins' in Hutton's theory of human history. The basic principle of this latter theory is the idea of progress. But Hutton was not concerned with the progress of the species from its place of origin, but of the individual nation wherever and whenever it occurred. Civilisations, also, are ephemeral, and they decline. So that although there is a progressive principle behind human history as against the strictly cyclical principle which we usually associate with Hutton's earth theory, there is a more fundamental similarity when the two historical realms are considered in the large: kingdoms and continents are but units for marking the processes of decay and renewal through which the form of the human species and of the whole earth are endlessly perpetuated.

Hutton was not concerned with absolute figures for the times involved. The life-span of even one continent was one of 'indefinite duration', and Hutton was content with letting the form of his theory make whatever demands it wanted upon time and then denying on epistemological grounds that any limit could justifiably be placed on these demands being met. He made, however, some suggestive remarks concerning the time through which the human species has existed. In one place he noted that, according to 'written history', the origin of man is 'little removed' from the present, while natural history provides no further evidence of a 'high antiquity' for the human race, in contrast with the fossil evidence for the existence of the lower animals at a time 'extremely remote'. (24) Elsewhere, he referred to the existence of the human species through a 'succession of ages not to be numbered', (25) while again his vision of the unending vicissitudes of nature and society led him to refer to the earth as 'a system calculated for millions, not of years only, nor of the ages of man, but of the races of men, and the succession of empires'. (26)

At a general level, there is some comparison between the history of the earth and the history of man as Hutton conceived of them, but we must note the essential contrast. Although Hutton of course recognised that there was some sense in which human societies are 'physically' determined, he treated this as a lesser factor: the life of the nation is the life of its human spirit, and the spiritual health of the nation is the main prerequisite for its survival. This is also the level at which nations are, through time, bonded each to each in a continuing 'succession', in the sense of a transcultural passing on of tradition such as that which Hutton thought had passed through the Egyptians to the Greeks, then to the Romans and on to the philosophers of modern Europe.

That perennial wisdom to which every new society must rise was, in Hutton's view, severely distorted by the form of Christianity represented by Burnet. The latter's <u>Sacred theory</u>, said Hutton, 'surely cannot be considered in any other light than as a dream, formed upon the poetic fiction of a golden age, and that of iron which has succeeded it'. (27) He added that at first sight the theory may seem empirically sound but this would be a misperception caused by 'a partial view of things'. Hutton's commitment to the idea of moral progress set him hard against the idea of primal purity followed by degeneration, which was too caught up in the Christian idea of the fall of man against which his own religious philosophy was a systematic protest. Along with his rejection of the fall and of the miracle of resurrection, Hutton rejected the 'partial view' of nature which they entail, that is the particular moulding of the idea of nature which they together effect for the purpose of their own support. But as we have seen, Hutton was merely opposing Burnet's scheme with his own which was analogous in structure. He systematically opposed his ideas of progress and immortality of the soul to Burnet's ideas of the fall and bodily resurrection, and his whole teleology of nature served him in this confrontation.

I have dealt so far only with the history of the earth and the history of man, without taking up the question of the history of life in general. This last question does have a bearing on our central issue of the afterlife, and I intend to discuss this with particular reference to Lyell. Lyell was forced to defend the doctrine of progressive immortality, to which he subscribed, under the pressure of new circumstances in the geological sciences.

The respective geologies of Hutton and Lyell are very different on the biological side, because it was only after Hutton's death that the phenomenon of extinction was made certain and that the idea of a 'progression' in the succession of the forms of life contained in the fossil record arose.

Hutton had been concerned with fossils primarily as signs of the marine origin of strata, and although he was aware of the fact that strata could be characterised by their faunal assemblages, this was an awareness quite outside the context of systematic stratigraphic palaeontology. Hutton's geology was not 'historical geology' in the modern sense; biology was integral to his theory of the earth for quite other reasons. Dealing with the earth always <u>qua</u> habitable earth, with conditions of habitability being maintained only on balance across a considerable geological time-span and under circumstances of considerable geological change, Hutton was interested in the provision, on the biological side, of ecological harmony between plants and animals and their geological environment.

Correlated with Hutton's steady state dynamism in earth theory, was his idea of the ultimate preservation of organic forms at the species level. Hooykaas has demonstrated the parallel between the history of life and the history of the earth in Hutton's work, noting that in each case, his uniformitarian principles were such as to foreclose any possibility of historical 'progression'. (28) What I want to discuss here is the causal relation between the two realms. In order to understand the sense in which in Hutton's system of nature the geological environment is adapted to the purpose of maintaining 'a system of living bodies perpetuating their species', (29) we must turn to his ideas on biological variation and on the selection of favoured varieties in what he refers to as the 'struggles of animated nature'. (30) His own agricultural practice left him well acquainted with the facts of variation and of artificial selection, and he argued that under natural conditions certain varieties would be favoured in the struggle of organisms to survive in a changing geological environment. A species would be fitted to survive under conditions of change due to what Hutton called its inherent 'principle of seminal variation' which, following the action of environmental selection upon the varieties thus formed, leads to the distinction of intra-specific races. (31) Variation and selection are purely intra-specific phenomena, by analogy with the ultimate constraining of geological change within an overall steady state pattern, or of 'occasional' or 'aberrant' motion within a planetary orbit. Hutton therefore did not believe that varieties could accumulate in one direction, thus leading the variety beyond the limits set by an original specific type. Were variation to become an end in itself, rather than just a means of preservation of the species, the result would be ecological disharmony and the 'degeneration' of the species. Hutton thought variation was limited, and he certainly did not envisage the rise of new species in this way.

Hutton had not been faced with the facts of extinction and of the appearance of new organic forms in the geological record, nor with the further idea that this was a record of the 'progression' of forms, nor of course with the idea of evolution. However, he was acquainted with the idea of organic transformation, based on the idea that the 'scale of beings' was a temporalised scale. Hutton believed in the scale of being only in the sense of a static hierarchy.

One of the works which Hutton had reviewed in chapter three of the _Theory_ was De Maillet's _Telliamed_, which he commended to the extent that it 'has something in it like a regular system such as we might expect to find in nature', while adding that it is yet 'only a physical romance, although apparently better founded than most of that which has been written on the subject'. (32) One of his principal objections becomes more apparent when we turn to his discussion of the formation of intra-specific races by selection, to which he added, 'we are not here to indulge the romantic fancy of a Telliamed, forming fowls of flying fish, and men of mermaids or some aquatic animal'. (33) Hutton did not feel the need to argue the point further in this context where the issue was 'philosophical materialism'. For a materialist philosophy of man was what Hutton, in the _Investigation of the principles of knowledge_, was mainly concerned to refute. As we have seen, questions as to the origin of the earth, and of the historical origin of the human species, were not entertained by Hutton as legitimate philosophical problems, and he took the same view regarding speculation on the origin of the individual soul. His philosophy in fact takes as its starting point the individual consciousness, from whence it proceeds to map out a schema for the realisation of its intuited capacities - and in so far as it deals with the problems of materialist reduction of the psychological world, it does this by statements of principle. It was Hutton's conviction that the human mind cannot get beneath or behind itself in order to explain the origins of either thought or language in some thing other than itself. Hutton did not argue the point against a background of speculation on the historical origins of the human species in the sense of a transformation of a different, 'lower', animal order. The issues with which he was engaged, those arising within the context of physiological psychology and of speculation on the origins of human language in the cries of the beast, were, as he perceived them, without any developed context of a truly historicist nature, and he could accordingly dismiss a writer like De Maillet without further ado.

The situation was very different from Lyell, however. Lyell considered that man's progressive nature marked him distinct in kind from the rest of the animal creation, but he saw that this idea could be deeply compromised by the further idea of a progression in the forms of life during geological time, with man placed at the end of the series. For although progression could be understood as a succession of special creations, thus in intent preserving man's autonomy, special creation was an opaque notion which could be replaced with a naturalistic theory (Lamarkian transmutation, for example) applying equally to man as to the other animals.

Lyell began by denying, in the _Principles of geology_ (1830-3), that there was a sound case for progression in the fossil record, alleging that it arose from the imposition of extraneous ideas on the geological data, either scriptural ideas of a successive creation, or transformationist ideas with an origin largely in non-geological biology. He thought the latter were merely 'ingrafted' upon the fossil record, marshalling the evidence behind the image of progression. It was a move supported, he also thought, by the explanation of the phenomenon of extinction which it provided. Thus one of Lyell's moves at this stage was to provide an alternative explanation of extinction, thus rendering transmutation gratuitous in this context and thereby exploding the image of progression which it generated. Lyell's own explanation was framed according to his uniformitarian principles, making extinction of species an integral part of his steady state system of the earth.

Lyell did not continue in his attempt to suppress the idea of progression, although he did continue to specify certain necessary qualifications to it. In 1863, in The antiquity of man, Lyell accepted both progression and transmutation as reasonable and useful geological hypotheses, while attempting to maintain the autonomy of the idea of progression in so far as it was to be used in the characterisation of human nature. Although man may be linked with the rest of animal nature in respect of his physical organisation, his moral and religious sense, and his 'power of progressive and improveable reason', marked him as distinct. And because it was a distinction in kind it could not, as Lyell pointed out in his Scientific journals of 1856, be considered as but one more step in a progression based on changes in physical organisation:

> A law of progressive developm.[t] is the only way of reconciling the successive appearance of higher and higher beings on the stage, but can any such law evolve the rational out of the irrational. It is here that the analogy fails. (34)

But when Lyell turned to a more positive appraisal of the importance of progression and transmutation for the problems of the origin and future of man, his ideas did not take a very definite form. For one thing, when considering the question of the origin of the human soul in the instincts of the lower animals, he proceeded to generalise it: the origin of the soul of man in the geological past is no more of a problem than the appearance of rationality in the human infant at the present day, and the extinction in the past of forms transitional between brute and human is no more of a problem than that of still birth or premature death at the present day. Lyell, in other words, threw the problem in the direction of traditional theological debate on the origin of the soul. Similarly, when appraising the problem of immortality, Lyell was unsure of the implications of the doctrine of geological progression. It must be recognised that Lyell did at times subscribe to notions about the grounds of religious belief which would, if followed exclusively, mean that there were no such implications; as he noted on the 'instinct of future state':

> Those aspirations which spring not from the inductive inferences of the Reason, but from some provision in Man's moral Nature, is [a] human instinct higher & more certain than either revelation or external sources of information. (35)

Lyell was prepared, in other words, to let the soul be 'its own revelation', to quote William Greg whose Creeds of Christendom (1851), with its advocacy of immortality through spiritual progress as against bodily resurrection, was evidently of support to Lyell in his reflections on the soul.

However, Lyell still attempted, in a hesitant way, to bring his idea of immortality into line with the idea of progression. This he did by speculating on the analogy between the phenomenon of progression in general, and the progressive nature of the soul, seeking in the former an image of the 'endless variety and improvement' which marked the idea of progressive immortality. But Lyell did not pursue this line of speculation, nor might it have been profitable for him to do so, for the simple reason that, as already mentioned, contemporary speculation on progression did not treat with sufficient care the distinction between the moral and the physical, so that progression in nature was a dangerous ally for the progress of the soul. Instead, by 'taking the inner light & intuition as guiding Man to a belief in the immortality of the soul', (36) Lyell could interpret the data which were for other thinkers evidence for man's animal ancestry as but further evidence that the problem of the origin of the soul was a perennial enigma whose each new form was as inscrutable as the last.

Notes

1 J. Hutton, "Theory of the earth. With proofs and illustrations", 2 vols., Edinburgh, 1795.

2 J. Hutton, "Dissertations on different subjects in natural philosophy", Edinburgh, 1792; idem, "An investigation of the principles of knowledge, and of the progress of reason, from sense to science and philosophy", 3 vols., Edinburgh, 1794.

3 P. Gerstner, 'James Hutton's theory of the earth and his theory of matter', Isis, 1968, lix, 26-31; P. M. Heimann & J. E. McGuire, 'Newtonian forces and Lockean powers: concepts of matter in eighteenth-century thought', Hist. Stud. Phys. Sci., 1971, iii, 233-306; P. M. Heimann, '"Nature is a perpetual worker": Newton's aether and eighteenth-century natural philosophy', Ambix, 1973, xx, 1-25; R. S. Porter, "The making of geology; earth science in Britain, 1660-1815", Cambridge, 1977.

4 Heimann & McGuire, op.cit. (3).

5 E.g. D. Dean, 'James Hutton on religion and geology: the unpublished preface to his Theory of the Earth (1788)', Ann. Sci., 1975, xxxii, 187-93; R. Hooykaas, "The principle of uniformity in geology, biology and theology", Leiden, 1963, pp.180ff.

6 Hutton, op.cit. (1), i, 275.
7 Ibid., 284.
8 Ibid., 296.
9 Ibid., 146.
10 Cf. note (7).
11 Hutton, op.cit. (1), i, 187.
12 Ibid., 11-12.
13 Cf. note (8).
14 Hutton, "Investigation", op.cit. (2), iii, 302.
15 Ibid., 721.
16 Ibid., 200.
17 Hutton, op.cit. (1), i, 12.
18 Ibid., ii, 257.
19 Ibid., i, 200.
20 Ibid., ii, 239.
21 Ibid., i, 270.
22 C. Lyell, "Principles of geology", 3 vols., London, 1830-3, i, 15, 14n.
23 Ibid., i, 64.
24 Hutton, op.cit. (1), i, 18-19.
25 Hutton, "Investigation", op.cit. (2), iii, 85.
26 Hutton, op.cit. (1), i, 372.
27 Ibid., i, 271.
28 R. Hooykaas, 'The parallel between the history of the earth and the history of the animal world', Arch. Int. Hist. Sci., 1957, x, 3-18.
29 Hutton, op.cit. (1), i, 280.
30 Hutton, "Investigation", op.cit. (2), i, 9.
31 Ibid., ii, 503.
32 Hutton, op.cit. (1), i, 271.
33 Hutton, "Investigation", op.cit. (2), ii, 500.
34 C. Lyell, "Scientific journals on the species question", ed. L. G. Wilson, New Haven & London, 1970, p.87.
35 Ibid., p.348.
36 Ibid., p.382.

The natural theology of the geologists: some theological strata

JOHN HEDLEY BROOKE

> I have long ago found out how little I can discover about God's absolute love, or absolute righteousness, from a universe in which everything is eternally <u>eating</u> everything else . . . The study of nature can teach no <u>moral</u> <u>theology</u>. It may unteach it, if the roots of moral theology be not already healthy and deep in the mind. (1) [Charles Kingsley]

A convenient point of departure for any discussion of natural theology is the ambiguity which arises when one refers to the apologetic strategy of its exponents. The clerical geologists of the nineteenth century, when they reconstructed their Creator from His Footprints, might have been defending their faith, their science, or both. Consequently, in the classification of the various functions which arguments from design were called upon to fulfil, it is surely wise to distinguish between those that were intrinsic to particular theological traditions and those more extrinsic, social and mediating functions associated with the promotion of science itself. My object in this essay is to clarify the different functions which appeals to design could fulfil, in order to suggest that further ambiguities associated with the propositions of natural theology allowed those propositions to serve a socially diplomatic purpose during a period when religious deviation had, or was certainly seen to have, social consequences. It would not follow from this that when a Buckland, a Sedgwick, or even a Lyell, publicised the evidence for a prescient God, he was being disingenuous. It would, however, follow that the gradual disappearance of references to the Creator from popular scientific works could not adequately be explained without exploring changing attitudes to religious dissent.

If my emphasis falls on the mediating functions of natural theology, it is most certainly not intended to conceal or demean those appeals to design which properly belong to the theological enterprise. The most rudimentary attempt to classify the functions of natural theology would have to recognise that design arguments have contributed to the erection of systematic theologies in which their status and value were explicitly discussed. (2) More urgently, arguments drawn from nature have served to engage the infidel and to expose the inconsistency of his presuppositions. (3) In addition, they have been invested with a preparatory role, that of informing the tactics of pre-evangelism. (4) Nor can one neglect the prominent unifying role of natural theology for those whose vision of nature was conditioned both by scientific and religious concerns. The value of arguments from design in achieving an integrated outlook is reflected in Buckland's definition of geological knowledge as 'the knowledge of the rich ingredients with which God has stored the earth beforehand, when He created it for the then future use and comfort of man', (5) as it is also in Sedgwick's quest for 'sermons in stones'. (6) Arguments from design could also perform a therapeutic role in reassuring one's brethren whose doubts might have been occasioned by the latest science. Ten thousand creative acts, Sedgwick assured a friend, had been recorded on stony tablets. (7)

If design arguments helped to reassure the faithful, they also served to reassure oneself. In his study of *Belief and unbelief since 1850*, H. G. Wood emphasised that confidence in the objective character of ethical distinctions and standards was to become the 'sheet-anchor of many Victorians when their religious beliefs were shaken or obscured'. (8) His point was that the Victorians had a way of finding their way back to faith 'through their loyalty to their moral convictions'. (9) Commitment to the existence of design in nature had, albeit to a lesser extent, surely fulfilled a similar function for those whose Christian beliefs were progressively overlaid by the conclusions of their science - conclusions associated with the sheer enormity of space and time which could exert subtle psychological pressures transcending the more specific problems raised by an encroaching uniformity of nature or the questioning of biblical authority. (10) The preoccupation with coal, so common in the natural theology of the 1820s and 1830s, certainly reflects the providentialism of the day, but it may well also reflect the need for reassurance that man had not been forgotten before he had existed. (11) Furthermore, there is no doubt that discussions of natural theology could be used to instruct the uninitiated in the safest apologetic stance, at the same time reproaching those who were bent on preserving a more vulnerable position. 'What are we to think of those officious penmen', Baden Powell was asked by Richard Owen, 'who go about alarming the well meaning sheep of their flock and the ignorant of all kinds, by telling them virtually that they must give up God if so and so in science be demonstrated?' (12) The fact that in this same letter Owen referred to a 'Puseyite reptile' who had crossed his path, is a reminder that this edifying function of natural theology cannot be detached from the ecclesiastical polemics of the day. (13)

Any discussion of design arguments which threatened to override these intrinsically religious functions would lead to oversimplification. Indeed, failure to take them properly into account can lead to distortion in at least three respects. There is the distortion bred by the failure to distinguish a theological from a scientific point as appears to have occurred in commentaries on Richard Owen's attitude to the transmutation of species: sympathetic references to a continuity of secondary causes have been construed as evidence of a positive commitment to 'evolution', (14) whereas the context in which Owen made such references suggests their theological rather than their scientific import. (15) He was anxious to correct those who wrote as if theology should retain a vested interest in the scientifically inexplicable, his efforts in that direction being in no way inconsistent with his habit of criticising the latest speculations on speciation. (16) Secondly, there is the distortion born of the failure to relate contending schemes of natural theology to the respective religious traditions of which they can be an expression. Fortunately, there has been a growing recognition that pre-Darwinian natural theology cannot be reduced to the system of Paley, (17) and that competing forms of natural theology have structured competing interpretations of the geological past. (18) I have insisted elsewhere that to understand the plurality of worlds debate between Brewster and Whewell it is necessary to appreciate the respects in which the natural theology of a Scottish evangelical and that of a liberal Anglican could be so divergent as to become almost mutually exclusive. (19) Failure to grant the importance and continuity of these theological functions would lead to a third distortion in the context of assessing the impact of Darwin, since the same demands remained pressing enough to ensure the survival of duly modified arguments long after Darwin's explanatory programme had embarrassed the common-sense teleology of Paley.

Having stressed that it is not my purpose to subjugate the theological functions of natural theology to the mediatorial, it is still appropriate to question the status accorded to design arguments in the theological literature of the day. For it is striking how, even in the 1830s and 1840s, the status granted to the arguments of natural theology was frequently subordinate to higher theological concerns. It requires only a minimal acquaintance with the works of Thomas Chalmers, Nicholas Wiseman, R. W. Church, even Sedgwick and Whewell, to appreciate how well the limitations of natural theology were recognised by Evangelical, Catholic, High Anglican, and Liberal Anglican spokesmen. (20) Questions affecting God's relationship to man, the authority of the Bible, Christology, the Resurrection, and the existence of a moral order, were usually of more central interest than the models of divine activity proffered by the geologists. (21) This is not to deny that specific issues were seen to have a direct bearing on the explication of the doctrine of providence, yet even among the clerical scientists there are explicit qualifications to the religious value of appeals to design. Whewell's qualifications are well known, (22) but they are there in Sedgwick too. Considerations drawn from nature 'fill the mind with feelings of the vastness of the power and skill employed in the mechanism of the world; yet, of the great Architect himself and of the materials employed by him, they give no adequate notion whatsoever'. (23) The conclusions of natural religion were of 'no small' moral worth, but even that required the proviso that they be kept 'in their proper place, and in subordination to truths of a higher kind'. (24) If the christian seeks the truths of natural religion it is not because he makes them the foundation of his faith, 'but because he believes'. (25) The contemplation of God through the wonders of the created world was a habit compatible with, but not constitutive of, firm religious belief. Indeed, the chief ground on which Sedgwick urged the study of natural religion was that it constituted 'a wholesome exercise for the understanding' - entirely appropriate for the young men of Cambridge in the 1830s. (26) Sedgwick approved Paley's Natural theology, but in the last analysis because it produced a 'cheerful sobriety of thought'. (27) It almost evaporated into the spirit in which the natural sciences should be pursued. (28)

Given these qualifications attached to the intrinsic, theological value of design arguments, and in the writings of someone like Sedgwick for whom their unifying role was so prominent, there is certainly a case for exploring the other functions which such arguments could perform. My suggestion is that the arguments of natural theology frequently performed a mediating function between different religious traditions, at a time when religious differences were associated with marked social and political consequences. Design arguments were well equipped to fulfil this mediating role precisely because they were doctrinally so imprecise. Historically, philosophically, and theologically they were ambiguous. By their very ambiguity they enabled a man of science to breathe freely, to defend his science, to defend himself if necessary, and to do so with integrity. On this hypothesis, their longevity is to be attributed to the fact that they could be affirmed both usefully and sincerely, their eventual demise to the fact that their intellectual hold was lost at more or less the same time that their diplomatic functions lost some of their urgency.

Arguments from design were historically ambiguous since they had been invoked by christians and deists alike. Indeed, historians have often observed how pressure from freethinkers and deists had the effect of steering christian apologetics towards presuppositions perilously close to those under attack. (29) In Butler's Analogy of religion, so popular again in the 1830s and 1840s, the question at issue had not been the validity of natural religion but its sufficiency. Arguments from design had been affirmed by Voltaire before they were affirmed by Paley.

Design arguments were also philosophically ambiguous in the sense that they had been promoted as part and parcel of philosophies of nature that were in every other respect at variance. One has only to recall the Leibniz-Clarke correspondence in which both protagonists argue for the wisdom of God but on the basis of apparently incommensurable precepts. (30) The temptation to reduce that particular debate to a confrontation between the workday God and the God of the sabbath, between theism and deism, has to be resisted. While it is tempting to see a deist emphasis in the Leibnizian appeal to divine foresight, and a theist emphasis in Clarke's return to divine omnipotence, their encounter really shows up the vagueness in such terms as 'deism' and 'theism'. It was Leibniz who stressed God's continual 'production' of nature, (31) Clarke who, in defiance of that formula, insisted that 'God discerns things . . . not by producing them continually (for he rests now from his work of creation) but by being continually omnipresent to everything which he created at the beginning'. (32) In this context, it was Clarke's God who was taking a rest.

The affirmation of design in nature was clearly compatible with a spectrum of deistic and theistic positions, from an extreme deistic model, which subsumed divine activity under natural laws, to a full biblical theism. Moreover, deists and theists alike could find themselves sharing virtually the same model for divine activity in the physical world. Consequently, without additional contextual clarification, it was (and for the historian still is) difficult to infer a man's theological stance from his remarks about divine wisdom. The difficulty is transparent in the works of James Hutton who combined an emphasis on final causes with a denial of the 'possibility of anything happening preternaturally or contrary to the common course of things'. (33) At the same time, he specifically dismissed the idea that 'the author of nature was limited by natural things' and proclaimed that 'every existence is to be resolved . . . to that infinite Being and superintending mind'. (34) Convergence between deistic and theistic philosophies of nature created a situation in which Sedgwick smote Vestiges with presuppositions curiously similar to those of his opponent. (35) As a matter of fact, Broad Churchmen were often thought to be deists reincarnate. (36)

There is a third, and arguably more important, sense in which design arguments were ambiguous: because they did not commit one to a specific doctrinal position, they could mediate between contending theologies when it was desirable to do so. For the student of nature, anxious to remove any suspicion in which his science might be held, an emphasis on design was an obvious way of pacifying as large an audience as possible. 'No rational theologian', wrote John Bird Sumner, 'will direct his hostility against any theory which, acknowledging the agency of the Creator, only attempts to point out the secondary instruments he has employed'. (37) When the student of nature was Buckland, such a licence from a respected divine was not to pass unacknowledged. (38) I am not suggesting that an emphasis on the agency of the Creator was a matter of expediency alone, but for those who wished to minimise the consequences of dissent, the mediating role of natural theology had an intrinsic appeal. One recalls the profound interest which Lyell showed in 'the effect which it would have here in modifying religious intolerance, if the social and educational equality of the different sects were greater'. (39) One also recalls Sedgwick's boast, from his prebendal stall at Norwich, that he had contrived 'to bring together more heretics and schismatics within my walls than ever had been seen before . . . since the foundation of the Cathedral'. (40)

The suggestion that natural theology flourished when it had mediating functions to perform derives some support from the context in which arguments from design were first given new prominence. In Restoration England, having been alarmed by the proliferation of puritan sects which was fast choking the credibility of the christian faith, Robert Boyle was attracted by the mechanical philosophy as a means of establishing a core of religious belief that would help to bind men together. (41) In a familiar passage from his _History of the Royal Society_, Thomas Sprat also complained of 'a mischief by which the greatness of the English is suppress'd', namely 'a want of union of interests, and affections . . . heightened by our civil differences, and religious distractions'. (42) The remedy, for Sprat, was not more spiritual controversy, but the calm study of nature. It may be spurious to imply parallels between England in the 1660s and in the 1830s; and yet Whewell in his _Bridgewater treatise_ explicitly aligned himself with a long intellectual tradition originating with Boyle, (43) while William Vernon Harcourt, in his 1839 address to the British Association, drew on a passage from Sprat. (44) Time, Harcourt added, changed circumstances but 'seems to have little effect in removing prejudice'. (45) A new scientific association, conscious of its public image, committed to removing prejudice, committed also to providing a sanctuary in which men of disparate beliefs could unite in the pursuit of their common interest - here was the perfect environment for design arguments to flourish. They flourished in Sedgwick's address before the 1833 meeting when he specifically warned that 'if we transgress our proper boundaries, go in to provinces not belonging to us, and open a door of communication with the dreary wild of politics, that instant will the foul Daemon of discord find his way into our Eden of Philosophy'. (46) Is it not arguable that an alliance between natural theology and the preservation of Sedgwick's Eden was reinforced in the ensuing decades when so many political issues became bound up with Church-Chapel confrontation and the 'gradual disestablishment' of the Church of England? (47) If the continual reflection of Church-Chapel confrontation in national politics 'succeeded in delaying for more than half a century the most obvious manifestations of secularisation', (48) it follows that the minimising of doctrinal differences and the defence of science against religious opposition were two functions of natural theology not immediately dispensable.

A letter from Lyell to Murchison, written in August 1829, indicates how Lyell had perceived the situation prior to the appearance of his _Principles of geology_:

> I trust I shall make my sketch of the progress of Geology popular. Old Fleming is frightened and thinks the age will not stand my anti-Mosaical conclusions and at least that the subject will for a time become unpopular and awkward for the clergy, but I am not afraid. I shall out with the whole but in as conciliatory a manner as possible . . . (49)

There is no indication here that Lyell was intimidated by thoughts of persecution. But there is a strong indication that those who wished to make their geology popular had to find the means of conciliation. Whether it was essentially a matter of good manners, as M. J. S. Rudwick has urged, (50) or whether there were more powerful pressures to conform, there can be little doubt that arguments from design, with their attendant ambiguities, provided just such a means.

The subject did, after all, become awkward for the clergy - for the clerical geologists as well as the anti-geological clerics. Sedgwick was harangued by the Reverend Henry Cole, (51) Buckland by more than his fair share of detractors. John Pye Smith came to the defence of the Geological Society when it was accused of subverting the fundamentals of revealed religion. (52) Almost every geologist of note had to endure the frustration of the importunate Dean of York. (53) Murchison was so vexed by the persistence of the Mosaic geologists that he made a point of approaching Harcourt as the ideal man to put them down. (54) When Harcourt duly obliged in his 1839 British Association address, it was with the arguments of natural theology that he struck the note of conciliation. The conclusions of human science, he proclaimed, were still in accordance with revealed religion, 'and none more remarkably than that which has been so falsely termed _irreligious_ geology; for as

Astronomy shows the unity of the Creator through the immensity of space, so does Geology, along the track of unnumbered ages, and through the successive birth of beings, still finding in all the uniform design of the same Almighty power . . . ' (55) In this way, the arguments of natural theology helped to defend the geologist's right to a method that could be pursued independently of narrow exegesis. (56)

Popular fears that geology was subversive of Scripture and, by implication, the stability of society, exerted a pressure all the more intense as the efforts of the German critics made their mark. It was with one eye on Germany that Pye Smith came to the rescue of Baden Powell, with whom he positively disagreed on the precise manner in which the concept of biblical accommodation should be applied: though Powell 'has unfortunately adopted the term "mythic poetry", he must not be understood as symbolizing with the men who have filled Germany. (57) Pye Smith is a striking example of someone who was deeply conscious of his mediating role. 'Perhaps', he surmised, 'the remarks of a mere <u>Christian</u> observer might, so far as they are of value, be received with less suspicion than those of a professedly scientific man'. (58)

It is the word 'suspicion' that recurs in the 1830s and 1840s. When Buckland was nominated Dean of Westminster, his wife wrote to Sir Philip Egerton:

> I think Sir R. Peel has shown much moral courage in making choice of a person of science, for it was sure to raise a clamour, and among good people too. It has always been quite unintelligible to me how it happens that on the Continent, where there is far less religion than in England, a man who cultivates Natural History, who studies only the works of his Maker, is highly considered . . . while, on the contrary, in England, a man who pursues science to a religious end (even who writes a <u>Bridgewater Treatise</u>) is looked upon with suspicion, and . . . contempt. (59)

For Buckland who, it is said, carried his hammer up his sleeve 'in order not to shock the feelings of the Scotchmen on Sunday', (60) for Sedgwick who made it plain that he preferred hammer and chair to whist and stall, (61) the arguments of natural theology helped to avert suspicion. 'I trust the time has now passed away', said Charles Daubeny in 1856, 'when studies such as those we recommend lie under the imputation of fostering sentiments inimical to religion', but since this was in the context of promoting science at Oxford, he took the precaution of invoking the 'Great Lawgiver of the universe' to make absolutely sure. (62) By their very ambiguity, design arguments afforded both freedom of manoeuvre and the means of pacification. In his letter to F. D. Maurice, from which I have taken my caption, Charles Kingsley made it clear that, in his opinion, the study of nature could actually unteach moral theology. He went on to say that he had hinted as much in Glaucus - but only hinted. 'I would do no more, because many readers mean by "moral" and "theology" something quite different from what you and I do, and would have interpreted it into a mere iteration of the old lie that science is dangerous to orthodoxy.' (63) Natural theology, in other words, was protected despite its deficiencies. Sustained by its mediating power, it continued to be sustained by Kingsley himself. (64)

A related concern for the weaker brethren provides an illustration of the kind of social pressure that encouraged recourse to natural theology - in this case to clear an individual as much as his science. In the late 1840s, Richard Owen was treated to a thinly veiled charge of atheism by the Manchester Spectator. The allegation was based on the assertion that Owen had adopted the development hypothesis of Vestiges. Owen's anger was roused not only because he had been misrepresented, but also, as he explained to Powell, because his 'worthy Lancashire relatives and friends' were beginning to feel 'uncomfortable'. (65) He retaliated with a stern letter to the editor in which he insisted that his recognition of an 'Ideal exemplar for the vertebrated animals, proves that the knowledge of such a being as man must have existed before man appeared'. (66) Design arguments were displayed to vindicate himself and to reproach his critic. (67) Provided he subscribed to the formula that the wisdom of God was manifest in the works of creation, a man of science could breathe freely in the company of colleagues with whom he might share a certain adventurousness of thought, and in the company of his relatives who exuded orthodoxy. (68) There is no reason to regard Owen's argument as a mere formality; at the same time it undoubtedly served its purpose.

I have suggested that it was the ambiguities associated with design arguments that made them attractive as mediating agents. Walter Houghton has observed that we are apt to forget the extent to which the Victorians, especially in the area of religion, 'adopted an equivocal position irrespective of what was proper or prudent'. (69) Even the theists, he added, were caught in an impossible dilemma: 'to be honest involved the guilty sense of perhaps undermining the moral foundations of society or depriving people of their spiritual comforts; to be reticent was to feel like a hypocrite'. (70) It seems to me that for the man of science who knew that his results would, however gently, shake the orthodox, the affirmation of design in nature provided an escape from the dilemma. One could enjoy the relief of sincerity and the virtue of conformity.

There is justification for this interpretation in the degree to which a certain equivocation is apparent in the literature. At the most rudimentary level, a sense of humour could inject ambiguity where superficially there was none. After hearing Buckland's tour de force on the Kirkdale cave, Lyell complained to Mantell that 'Buckland in his usual style enlarged on the marvel with such a strange mixture of the humorous and the serious that we could none of us discern how far he believed himself what he said'. (71) A degree of equivocation was also achieved by blending certain of Joseph Butler's propositions with the assertion that it was not improper to search out a mechanism for the origin of species. (72) This was Charles Babbage's strategy when he tried to pretend that his own concept of programmed miracle had episcopal blessing. (73) The fact is that a need for ambiguous words was felt by at least some of the geologists. The use of a word such as 'creation' sometimes reminds one of the 'weasel words' in our advertising industry. In this case, it was conveniently left open whether 'creation' was by divine fiat, by secondary causes, or by an appropriate mixture. 'In regard to the origination of new species', Lyell wrote to Herschel in 1836, 'I am very glad to find that you think it probable that it may be carried on through the intervention of intermediate causes. I left this rather to be inferred, not thinking it worthwhile to offend a certain class of persons . . . ' (74)

An engaging example of this ambiguity actually at work occurs in Pye Smith's exposition of the following statement from Powell:

> The sure monuments which we derive from the study of organic remains, disclose to us evidences of a series of gradual changes and repeated creative processes, going on without any one sudden universal intervention or creation of the existing world out of the ruins of a former. (75)

It should be noted how Pye Smith gave Powell the benefit of the doubt: 'The Professor, when he speaks of "creative processes" unquestionably understands interventions of the Deity, direct acts of power essentially different from any kind of generative evolutions.' (76) Knowing as much as we do of Powell's philosophy of nature, Pye Smith's confidence was surely misplaced. (77) Ambiguities of this kind continued to serve their purpose until Darwin, obligingly or not, set up his antithesis between special creation and natural selection in the final chapter of his Origin of species. And even Darwin, accusing himself of 'truckling', made a point of retaining the same useful ambiguity when the issue was not the origin of species but the origin of life. (78)

Given the operation of ambiguities at this level, it does not seem unreasonable to see design arguments working in a similar way but on a different level. In fact they worked rather better. In his first edition, Pye Smith felt he had to reproach Lyell for a careless reference to 'admiration - strongly excited, when we contemplate the powers of insect life, in the creation of which nature has been so prodigal'. (79) Lyell's use of the word 'nature' meant that on this occasion Pye Smith was not prepared to give the benefit of the doubt. 'O, why did not his heart grow warm within him, and bound with joy, at the opportunity of doing some homage to the God of glorious majesty?' (80) But, by the second edition, Pye Smith was reproaching himself. He was now confessing that he had not done justice to Lyell's own reference to the 'clear proofs of a Creative Intelligence, and of his foresight, wisdom and power'. (81) Arguments from design, once perceived, would mediate more successfully than ambiguous words such as 'creation'.

If design arguments were sustained, in part, by their social functions, there are certain moves one would expect to find in the natural theology literature. One would expect to find the occasional hint that references to divine wisdom were perhaps a shade too convenient. And there are such hints in Pye Smith's disdain of 'hypocritical affectation' (82) and in Lord Brougham's observation that 'scientific men were apt to regard the study of natural religion as little connected with philosophical pursuits . . . as a speculation built rather on fancy than on argument'. (83)

One would also expect to find an emphasis on natural theology whenever the interests of science were at stake. (84) One might even anticipate the occasional voice of doubt as to whether the moral virtues of a scientific education had not been overplayed. 'Perhaps, indeed', wrote the Duke of Argyll in 1855, 'like other zealous advocates, we may have sometimes overstrained our language . . . We cannot too earnestly disclaim the idea that the knowledge of physical laws can ever of itself form the groundwork of any active influence in morals or religion.'(85) A further expectation would be prominence given to design arguments when a man of science was on the defence. And the prominence is there: in Playfair's reprieving Hutton from the charge of atheism; (86) in the third volume of Lyell's _Principles of geology_, where Lyell scotched the rumour that his earth had existed from eternity; (87) and, as we have already seen, in the effort of Richard Owen to put his foot on the 'reptile' who had abused him.

If it was ambiguities associated with design arguments that helped to preserve them, one would imagine that those who presented their natural theology as an autonomous discipline would be anxious to reassure their christian audience that natural theology was, after all, a respectable part of Christianity. Precisely that strategy was pursued by Lord Brougham who devoted an entire chapter to appeasing those who were 'alarmed lest the progress of natural religion should prove dangerous to the acceptance of revealed'. (88)

It was in that context of appeasement that he declared it a vain and ignorant thing 'to suppose that natural theology is not necessary to the support of revelation'. (89)

Conversely, one might anticipate a sense of discomfort among the slightly less orthodox if natural theology were too overtly christianised. When Sir James Stephen was faced with Whewell's critique of a plurality of worlds, he experienced precisely that. Whewell appeared to be replacing a conventional natural theology with a purified, but far more oppressive, Christianity. If he was right, then the only world was this fallen world: 'if you are really right', Stephen responded, 'is it not simply impossible to adhere to our commonly receiv'd natural theology?' (90)

Finally, if the mediating functions of natural theology were of paramount importance, one would expect to find a degree of consternation among the scientific alumni when those functions were sabotaged. Historians have often commented on the fact that Vestiges sold like hot cakes whilst the scientific community pronounced it a recipe for disaster. The disparity is partly explained by the philosophical and scientific naivety of the book, partly by the fact that Chambers's theodicy cut straight across the argument from adaptation to design as it was understood bySedgwick or Whewell. But it is perhaps more fully explained by the fact that Vestiges represented a reductio ad absurdum of the mediating role of appeals to design. By launching such appeals in the context of a development hypothesis that smacked of both materialism and determinism, the author of Vestiges had sold the pass. (91) On this point, the perception of R. W. Church is as revealing as the ire of Sedgwick (92) or the judicious ambivalence of Herschel. (93) In his review for the Guardian, Church hit the nail on the head: 'The Vestiges warns us, if proof were required, of the vanity of those boasts which great men used to make, that science naturally led on to religion.' (94)

It may not be too far from the truth to say that the historiography of the relations between science and religion has been dominated by two contrary traditions - the one, rationalist in intent and having its roots in the trenches of White and Draper; (95) the other, recently dubbed 'revisionist', (96) exploiting the more fruitful interactions. A tendency in the former tradition has been to reduce the religious affirmations of eminent scientists to placatory gestures; a tendency in the other, perhaps, to minimise social pressures which were keenly felt. My own hunch is that design arguments outlived their rigour because they were sustained both by sincere religious belief and by ambiguities which catered for freedom of enquiry and for efforts at mediation.

It is an obvious consequence that attempts to understand the 'demise' of natural theology must not focus exclusively on the intellectual impact of Darwin's thesis. (97) Certainly the impact was great enough to make it difficult for some writing in the 1860s and 1870s to appreciate the sincerity of an earlier generation which had given so overt a religious meaning to biological adaptation. The beginnings of that transition are plain enough in the Lyell-Darwin correspondence where Lyell senses that his reluctance to go the whole way with Darwin is being attributed by Darwin himself to expediency. Lyell's feelings were hurt in two ways at once: 'My feelings . . . more than any thought about policy or expediency, prevent me from dogmatising as to the descent of man from the brutes.' (98) Certainly there are no grounds for minimising the impact of natural selection, or for discounting the sense in which, by the 1850s, the intellectual format of physico-theology had become self-defeating, as illustrated by the way the different responses to Vestiges were so much at variance. (99) Yet, at the same time, if design arguments had been sustained by their mediating function, their gradual disappearance from scientific works probably had something to do with a gradual relaxation of pressures which, in the 1830s and 1840s, had been more intense. The rejection of references to the Creator from scientific texts can hardly be attributed to Darwin alone, when Darwin himself was inclined to retain them. (100)

Robert Young has documented the way in which the conventional arguments of natural theology were, by the 1870s and 1880s, powerless to constitute a common intellectual culture. (101) A sensitive index of changing attitudes to natural theology is also provided by references to design in the Presidential addresses delivered before the British Association - sensitive because, as Lyell knew, critical comments on the clerical hold over education, however conciliatory in tone, were apt to be expurgated by the British Association Committee. (102) Of thirty-four main addresses between 1832 and 1865 no fewer than twenty-two made some reference to the wisdom of the Creator, and at least ten involved some development of the theme. (103) Between 1868 and 1878 the complexion of the addresses became a good deal more ruddy. Hooker, in 1868, referred to natural theology as 'that most dangerous of all two-edged weapons . . . a science falsely so called'. (104) It was possible, Hooker acknowledged, for the Reverend Dr Hanna to adduce a long list of eminent clergymen who had adorned science by their writings, but 'Dr Hanna omits to observe that the majority of these honoured contributors were not religious teachers in the ordinary sense of the term; nor does he tell us in what light many of their scientific writings were regarded by a large body of their brother clergymen'. (105) Three years later, William Thomson was expressing his conviction that the argument from design had been 'much lost sight of' in recent zoological speculation. (106) Three years later again, we arrive in Belfast where John Tyndall threw all caution to the winds. (107) 'How different is the position of matters in this respect in our day!' declared the President for 1877, Professor Allen Thomson. (108)

For Allen Thomson, the man of science, that enormous difference was simply due to the 'general progress of scientific knowledge', in which Darwin had the lion's share. (109) This is precisely the interpretation that would seem to be inadequate. In the 1830s and 1840s, design arguments had helped to dissociate science from fears of atheism and spasmodic fears of revolution. It was to the analogy between the fixed laws of nature and the fixed laws of society that Buckland returned for his Easter sermon in 1848: 'Equality of mind or body, or of worldly condition, is as inconsistent with the order of nature as with the moral laws of God.' (110) A fortnight earlier, every precaution had been taken to protect the Abbey from a potentially unruly Chartist meeting. (111) Design arguments had acted as mediating agents when religious deviation implied a loss of social status, as it did even for wealthy Unitarians. (112) 'If it were equally gentlemanlike', reported Lyell in 1847, 'men would take less and be lay teachers, so many of the best have scruples in undertaking the clerical office . . . if something could be done to place secular and clerical teachers on the same social footing, we might . . . be emancipated from our trammels . . . ' (113) But it had also been gentlemanlike not to disturb the faith of the orthodox and this was another pressure under which design arguments had been sustained. Sedgwick, for example, had a keen sense of 'practical wisdom', which meant that he was unhappy about the way in which F. D. Maurice and Sir James Stephen had spoken out on the duration of Hell. (114) It was not in good taste to 'disturb the congregation' - a sentiment shared in spirit by those who had valued a sense of decorum at the meetings of the British Association. In the latter context, the pressures for mediation were even more pronounced as a consequence of the desire to transcend doctrinal, denominational, and political differences. 'We are in no great risk', Whewell had affirmed in 1841, 'of deviating into literary, or metaphysical, or theological discussions'; (115) and yet, as if to eliminate the risk altogether, he immediately went on to recall the 'elevated strains of religious reflection' that had been heard from his predecessors! To harp on the themes of natural theology had been a way of avoiding discord.

It would obviously be misleading to suggest that all the geologists of the 1830s had experienced these pressures in the same way, but it might be less misleading to suggest that, by the late 1870s, the pressures themselves had relaxed. If that could be demonstrated, it would follow that the external functions which design arguments were, by then, less well equipped to perform were, in any case, no longer so necessary. To document the point in detail is beyond my scope, but some symptoms of the change are well known: the new opportunities for nonconformists to graduate from Oxford and Cambridge in the 1850s, the increased democratic involvement of dissenters in the 1860s, and the greater freedom of opinion open to clerics themselves, following the ultimate failure to impeach the authors of Essays and reviews. (116)

The change was gradual and certainly not an immediate consequence of scientific innovation. As late as 1865, Hooker wrote a telling letter to Darwin: 'It is all very well for Wallace to wonder at scientific men being afraid of saying what they think . . . Had he as many kind and good relations as I have, who would be grieved and pained to hear me say what I think, and had he children who would be placed in predicaments most detrimental to children's minds . . . he would not wonder so much.' (117) On the subject of the antiquity of man, for example, the balance was still so delicate in the early 1860s that Hugh Falconer could find relief in the reflection that it was he, and not the likes of Soapy Sam, who had exposed the modernity of an Abbeville jaw: 'had the expose been made by the enemy . . . the whole subject would have been put back quarter of a century'. (118)

The change was a gradual one, and it had begun well before 1859. In 1847, Lyell had informed George Ticknor that 'there is a move now in the right direction; but the clerical influence arrayed against all progressive sciences . . . is too powerful to be easily overcome'. (119) Four years later, Lyell was able to draw a specific contrast between the situation in 1851 and that which had obtained ten years earlier: 'several instances of theological works of free inquiry in an earnest spirit, printed by men who are suffered to retain professorships in Universities, although so outrageously unorthodox that ten years ago they would have been sent to Coventry in society . . . ' (120) A similar contrast was later drawn by Henry Sidgwick, this time between 1869 and 1849. Discussing the honest doubts of Arthur Hugh Clough, he remarked that 'his point of view and habit of mind are less singular in England in the year 1869 than they were in 1859, and much less than they were in 1849 . . . the opinions that we do hold we hold if not more loosely, at least more at arm's length'. (121) If it had once been ungentlemanly to go on the offensive, by 1874 Lyell considered it manly of Tyndall to have done so. (122)

There are signs that the mediating functions of natural theology were becoming more redundant despite the hostile reception accorded to Darwin's thesis. Early in 1862, Huxley was pleasantly surprised by the progress of liberal opinion. Of his Edinburgh lectures on man, he wrote to Darwin: 'Everybody prophesied I should be stoned and cast out of the city gate, but, on the contrary, I met with unmitigated applause!!' (123) By the 1870s, it has been argued, the open-mindedness which John Stuart Mill had so prized in his Essay on liberty, had been over-reached to the point of adopting a broad-minded toleration of all ideas regardless of their intrinsic merits: 'Instead of a genuine effort at mediation, men began to accept any convenient compromise, however loose and vague.' (124)

To reinforce the perspective I have tried to construct, I should like to conclude with reference to an interpretative framework supplied by a nonconformist minister writing in 1870, a minister who had been one of the first to graduate from University College, London and who lived to see his son take a first at Oxford. In his First principles of ecclesiastical truth, J. Baldwin Brown analysed what he considered to be the 'revolution of the last quarter of a century'. (125) Society had been tending towards the 'trial of the democratic principle'; (126) but, even excluding that political tendency, he was able to identify four components of his 'revolution'. Certainly, one of these four was the 'intellectual revolution', in which science had played a prominent role. Conceding that the 'deepest interest of the age is in science', he noted that among the doctors of science there had been the growth of a 'lofty and supreme indifference'. (127) Reflecting on the 'bitter and prolonged resistance which [theologians] have offered to every scientific doctrine', he was not even surprised. (128) The impact of science was part of the 'utter overthrow of ancient and venerated authority', part of a wider criticism of ideas and institutions on which the order of society had been believed to rest. (129)

But there were three other components to the revolution, and each of these arguably had a bearing on the fate of natural theology. Brown identified a social component: with the rise of commerce there had been a 'complete break-up and mixture of social orders'. (130) There had once been a 'sense of repose in a settled order, in fixed relations, duties and appointed work'. (131) Now nothing was fixed: there was no judge in Israel. The last quarter century had also seen an ecclesiastical revolution - the culmination of an increase in nonconformist churches and the concomitant decline in the 'principle of an established national church'. (132) The Establishment principle, he wrote, 'has at length grown so weak, that in the judgment of the Primate of England it has but a few years' lease of life'. (133) One recalls that the political disaffection created by the exclusive privileges of the Church of England had been abhorrent to Lyell. (134) As those privileges were reduced, so the need for mediation became less pronounced.

There was yet one more component - a theological revolution too. The prevalent Evangelical commitment to doctrinal unity had given way to a deeper sense of unity with respect to those beliefs which 'mould the life'. Departure from the minor matters of the accepted Creed, once the 'unpardonable sin in our most evangelical churches', was no longer so offensive. (135) In these days, Brown suggested, 'a man's theological opinions form a very bad measure indeed of his spiritual life'. (136) Where that belief was held, the mediating functions of natural theology were more manifestly redundant. 'Close and fruitful intercourse, through the railway, the steamboat, the electric telegraph, the penny post, the public meeting, and the cheap press' was now 'the notable feature of our modern civilisation.' (137) In short, religious toleration had increased: 'we seem to know all about each other'. (138)

A theological, as well as an intellectual, social, and ecclesiastical change had altered the environment in which the arguments of natural theology had been best fitted to survive. In fact, Brown's 'theological revolution' had implications more far-reaching still. It was a revolution that reflected 'the growing belief that the relation of God to the world is less fairly set forth by the relation of a king to his subjects, than by that of a father to his household'. (139) In a theological transition that was far from trivial, (140) the legislating God of physico-theology lost his kingdom as well as his power and glory.

Notes

1 Kingsley to F. D. Maurice, July 1856, in F. E. Kingsley (ed.), "Charles Kingsley: his letters and memories of his life", London, 1883, p.181.

2 See, for example, the systematic elements in the "Bridgewater treatise" of Thomas Chalmers: "On the power, wisdom, and goodness of God as manifested in the adaptation of external nature to the moral and intellectual constitution of man", 2 vols., London, 1833, ii, 282-93.

3 These were some of Chalmers's objectives in his _Astronomical discourses_ as he strove to draw the sting from those who had exploited the vastness of the universe and the plurality of worlds to perplex the faithful: D. Cairns, 'Thomas Chalmers' _Astronomical discourses_: a study in natural theology', Scot. J. Theol., 1956, ix, 410-21.

4 Thus, commenting on Boyle's assertion that 'natural religion is the foundation upon which revealed religion ought to be superstructed', one of William Buckland's critics was anxious to make it clear that 'Boyle's proposition . . . may refer only to the order in which the different truths of religion ought to be placed before the sceptic mind': J. Mellor Brown, "Reflections on geology, suggested by the perusal of Dr Buckland's Bridgewater treatise", London, 1838, p.5.

5 E. O. Gordon, "The life and correspondence of William Buckland D.D.", London, 1894, p.267.

6 Sedgwick to Canon Wodehouse, 7 February 1847, in J. W. Clark & T. Hughes, "The life and letters of the Reverend Adam Sedgwick", 2 vols., Cambridge, 1890, ii, 115. For Sedgwick's achievement of an integrated outlook, see the brief paraphrase by John Herschel of his performance on Tynemouth beach when the British Association met at Newcastle in 1838: ibid., i, 515-16.

7 Ibid., ii, 79-80. Similarly the anonymous essay of Whewell, "Of the plurality of worlds", London, 1853, was explicitly addressed to the perplexed believer.

8 H. G. Wood, "Belief and unbelief since 1850", Cambridge, 1955, chapter VI.

9 Ibid.

10 I have argued this point at greater length in my 'Natural theology and the plurality of worlds: observations on the Brewster - Whewell debate', Ann. Sci., 1977, xxxiv, 221-86. The point was well made by T. H. Huxley in a letter to Charles Kingsley: 'Whether astronomy and geology can or cannot be made to agree with the statements as to the matters of fact laid down in Genesis - whether the Gospels are historically true or not - are matters of comparatively small moment in the face of the impassable gulf between the anthropomorphism (however refined) of theology and the passionless impersonality of the unknown and unknowable which science shows everywhere underlying the thin veil of phenomena.' L. Huxley, "Life and letters of Thomas Henry Huxley", 3 vols., London, 1908, i, 345. See also Marcia Pointon's discussion of _Pegwell Bay_, by William Dyce, in 'Geology and landscape painting in nineteenth-century England', this volume.

11 See, for example, the author cited by John Pye Smith in his "The relation between the Holy Scriptures and some parts of geological science", London, 1854, p.434.

12 Owen to Baden Powell, 26 January 1850. I am indebted to Jonathan Hodge for a transcription of this letter.

13 Thus the virtues of natural theology were extolled by Sedgwick in the context of hammering the Mosaic geologists and of educating those 'excellent christian writers on the subject of geology' who had indulged, as he had once himself, in premature schemes of scriptural concordance: A. Sedgwick, "A discourse on the studies of the university", 1833; reprinted Leicester, 1969, pp.94, 107-8.

14 G. Basalla, W. Coleman, & R. H. Kargon (eds.), "Victorian science", New York, 1970, p.300, where it is said that Owen 'opposed Darwinian natural selection but not the idea of evolution'. The use of the word 'evolution' conveys an oversimplification since Owen's 1858 address to the British Association in fact included a critique of the idea of the transmutation of species (ibid., p.321), just as he had privately attacked _Vestiges_ when it had first appeared. On this latter point see J. H. Brooke 'Richard Owen, William Whewell and the "Vestiges"', Brit. J. Hist. Sci., 1977, x, 132-45.

15 In his analysis of the term 'distinct creation', Owen showed how the ever changing positions of land and sea disqualified the view that 'the Apteryx of New Zealand and the Red-grouse of England were distinct creations in and for those islands respectively' (Owen, Address to the British Association, 1858, reprinted in Basalla, Coleman, & Kargon, op.cit. (14), p.326). Yet the point on which he wished to insist was this: 'supposing both the fact and the whole process of the so-called "spontaneous generation" of a fruit-bearing tree, or of a fish, were scientifically demonstrated, we should still retain as strongly the idea, which is the chief of the "mode" or "group of ideas" we call "creation", viz. that the process was ordained by and had originated from an all-wise and powerful First Cause . . .' (ibid., pp.326-7).

16 For Owen's contrast between his own 'view of the origin of species by a continuously operative creational law' and the transmutation mechanisms proposed in Vestiges and in Darwin's Origin of species, see D. L. Hull, "Darwin and his critics", Cambridge, Mass., 1973, pp.171-2, 176, 184, 187.

17 P. J. Bowler, 'Darwinism and the argument from design: suggestions for a reevaluation', J. Hist. Biol., 1977, x, 29-43; idem, "Fossils and progress: paleontology and the idea of progressive evolution in the nineteenth century", New York, 1976, chapter III. Access to the literature on natural theology can also be gained from the following studies: W. F. Cannon, 'The problem of miracles in the 1830s', Victorian Stud., 1960, iv, 4-32; idem, 'The bases of Darwin's achievement: a revaluation', Victorian Stud., 1961, v, 109-34; C. C. Gillispie, "Genesis and geology", New York, 1959; R. Hooykaas, "Natural law and divine miracle: a historical critical study of the principle of uniformity in geology, biology and theology", Leiden, 1959; idem, 'The parallel between the history of the earth and the history of the animal world', Arch. Int. Hist. Sci., 1957, x, 3-18; J. C. Greene, "The death of Adam: evolution and its impact on western thought", Ames IA, 1959; idem, 'The Kuhnian paradigm and the Darwinian revolution in natural history', in D. H. D. Roller (ed.), "Perspectives in the history of science and technology", Norman, Okla., 1971; M. J. S. Rudwick, "The meaning of fossils", New York & London, 1972; R. M. Young, 'Darwin's metaphor: does nature select?', The Monist, 1971, lv, 442-503; idem, 'The historiographic and ideological contexts of the nineteenth-century debate on man's place in nature', in M. Teich and R. M. Young (eds.), "Changing perpectives in the history of science", London, 1973, pp.344-438.

18 See, for example, the contrast between Daubeny and Scrope, in M. J. S. Rudwick, 'Poulett Scrope on the volcanoes of Auvergne: Lyellian time and political economy', Brit. J. Hist. Sci., 1974, vii, 205-42.

19 Brooke, op.cit. (10).

20 For Chalmers, natural theology had a limited scope and within that scope there was a degree of subordination: 'the theology of nature sheds powerful light on the being of a God, and . . . can reach a considerable degree of probability, both for his moral and natural attributes . . . when it undertakes the question between God and man, that is what it finds to be impracticable' (Chalmers, op.cit. (2), p.285). See also D. F. Rice, 'Natural theology and the Scottish philosophy in the thought of Thomas Chalmers', Scot. J. Theol., 1971, xxiv, 23-46. Wiseman allowed the works of God to be a true but faint image of the Workman, merely a 'glass' through which Divine attributes could be contemplated (N. Wiseman, "Twelve lectures on the connexion between science and revealed religion", London, 1842, pp.240-1). Church dismissed Paley's apologia as a 'starveling' argument, in much the same spirit as Newman had already done: J. D. Yule, 'The impact of science on British religious thought in the second quarter of the nineteenth century', University of Cambridge PhD thesis, 1976, pp.266-8.

21 Cf. S. Best, "After thoughts on reading Dr Buckland's Bridgewater treatise", London, 1837, pp.1-2. For a similar emphasis in Newman's reaction to Darwin, see his letter to Pusey, 5 June 1870, in C. Dessain & T. Gornall (eds.), "The letters and diaries of John H. Newman", Oxford, 1973, xxv, 137-8.

22 W. Whewell, "Astronomy and general physics considered with reference to natural theology", London, 1833, chapter IX. In his essay on the plurality of worlds, Whewell insisted that analogies between the government of the natural and spiritual worlds were to be dwelt upon 'with great reserve and caution', op.cit. (7), 2nd edn., London, 1854, p.74.

23 Sedgwick, op.cit. (13), p.87.

24 Ibid., pp.87-8.

25 Ibid., p.92.

26 Ibid., p.88.

27 Ibid., p.88.

28 Ibid., p.108. If this should seem too drastic a paraphrase it is well to remember that for Sedgwick, as for Chalmers, the strongest argument that natural theology had to offer derived not from the universality of contrivance but from the universality of conscience; cf. Rice, op.cit. (20), p.35. Such qualifications are important because they have an obvious bearing on Owen Chadwick's recent remark that there is very little evidence, in the aftermath of the Darwinian debates, that churchmen were particularly worried by the loss of Paley's narrow teleology; O. Chadwick, "The secularization of the European mind in the nineteenth century", Cambridge, 1975, p.183.

29 Two recent examples are P. Gay, "The Enlightenment - an interpretation", 2 vols., London, 1970, ii, 142; and M. C. Jacob, "The Newtonians and the English Revolution", Hassocks, Sussex, 1976, p.171.

30 H. G. Alexander (ed.), "The Leibniz-Clarke correspondence", Manchester, 1956; F. E. L.Priestley, 'The Clarke-Leibniz controversy', in R. E. Butts & J. W. Davis (eds.), "The methodological heritage of Newton", Oxford, 1970, pp.34-56.

31 Leibniz's fourth paper, in Alexander, op.cit. (30), p.41.

32 Clarke's fourth reply, ibid., p.51.

33 J. Hutton, "An investigation of the principles of knowledge, and of the progress of reason, from sense to science and philosophy", 3 vols., Edinburgh, 1794, ii, 309.

34 Ibid., ii, 415-17. There are therefore grounds on which Hutton could be described as either deist or theist. On Hutton's philosophy of nature, I have benefited from a study by my colleague, Peter Heimann: 'Voluntarism and immanence: conceptions of nature in eighteenth-century thought', J. Hist. Ideas, 1978, xxxix, 271-83. See also, Grant, 'Hutton's theory of the earth', this volume.

35 Gillispie, op.cit. (17), pp.166-7.

36 M. A. Crowther, "Church embattled: religious controversy in mid-Victorian England", Newton Abbot, 1970, p.35.

37 J. B. Sumner, "A treatise on the records of creation", 2 vols., London, 1816, i, 283.

38 W. Buckland, "Vindiciae geologicae", Oxford, 1820, p.26.

39 K. M. Lyell, "Life, letters, and journals of Sir Charles Lyell", 2 vols., London, 1881, ii, 122; see also ii, 86, 168-9, 322.

40 Clark & Hughes, op.cit. (6), i, 437. As Paul Marston rightly pointed out to me, this should not be taken to imply that Sedgwick was attracted to his 'heretics and schismatics' because they could share his natural theology; he was attracted by their piety.

41 Boyle's sensitivity to the proliferation of sects is apparent in his private correspondence in 1646 and 1647. See P. M. Rattansi, 'The social interpretation of science in the seventeenth century', in P. Mathias (ed.), "Science and society, 1600-1900", Cambridge, 1972, pp.1-32 (21); also C. Webster, "The Great Instauration: science, medicine and reform, 1626-1660", London, 1975, pp.94-5; and J. R. Jacob, 'The ideological origins of Robert Boyle's natural philosophy', J. Eur. Stud., 1972, ii, 1-21 (3).

42 T. Sprat, "The history of the Royal Society", ed. J. I. Cope & H. W. Jones, London, 1959, pp.426-7.

43 Whewell, op.cit. (22), London, 1839, pp.354, 361-2.

44 W. Vernon Harcourt, in "Report of the . . . British Association for the Advancement of Science, 1839", London, 1840, p.16.

45 Ibid.

46 A. Sedgwick, in "Report of the . . . British Association for the Advancement of Science, 1833", London, 1834, p.xxix.

47 A. D. Gilbert, "Religion and society in industrial England", New York & London, 1976, pp.162-72, 205; G. Kitson Clark, "The making of Victorian England", Edinburgh, 1962; reprinted London, 1965, p.162.

48 Gilbert, op.cit. (47), p.205. Cf. A. Briggs, "The age of improvement", London, 1959, p.337.

49 Lyell to R. Murchison, 11 August 1829, Murchison papers, Geological Society.

50 M. J. S. Rudwick, 'Charles Lyell, FRS (1797-1875) and his London lectures on geology', Notes Roy. Soc. Lond., 1975, xxix, 231-263 (236, 244).

51 H. Cole, "Popular geology subversive of divine revelation", London, 1834.

52 Pye Smith, op.cit. (11), pp.456-9.

53 W. Cockburn, "The creation of the world, addressed to R. I. Murchison Esq., and dedicated to the Geological Society", London, 1840; idem, "The Bible defended against the British Association", 5th edn., London, 1845, pp.33-9.

54 For this point I am indebted to Jack Morrell.

55 Harcourt, op.cit. (44), p.20.

56 In his Bridgewater treatise, Buckland put it this way: 'If . . . geology should seem to require some little concession from the literal interpreter of Scripture, it may be fairly held to afford ample compensation for this demand by the large additions it has made to the evidences of natural religion . . . ' ("Geology and mineralogy considered with reference to natural theology", 2 vols., London, 1837, i, 14).

57 Pye Smith, op.cit. (11), p.442; cf. pp.454, 449.

58 Ibid., p.329.

59 Mary Buckland to Sir Philip Egerton, November 1845, in Gordon, op.cit. (5), p.219. Such contrasts, with France in particular, were commonplace in the propaganda of the Declinists. Whether they were just is, of course, another question.

60 Ibid., p.34.

61 Sedgwick to Canon Wodehouse, 12 October 1837, in Clark & Hughes, op.cit. (6), i, 497.

62 C. Daubeny, in "Report of the . . . British Association for the Advancement of Science, 1856", London, 1857, pp.lxxi-ii.

63 F. E. Kingsley, op.cit. (1), p.182.

64 I am thinking here of the widely quoted passage in which Kingsley, in fact, rather overdid the protection: 'We might accept all that Mr. Darwin, all that Prof. Huxley, all that other most able men have so learnedly and acutely written on physical science, and yet preserve our natural theology on the same footing as that on which Butler and Paley left it.' Quoted in J. Dillenberger, "Protestant thought and natural science", London, 1961, p.235.

65 Owen, op.cit. (12).

66 Manchester Spectator, 22 December 1849, p.4.

67 Ibid.; Owen's letter to the editor is reproduced, in full, in Brooke, op.cit. (14).

68 As Jonathan Hodge has written, with specific reference to Owen, 'saying that a common plan or a unity of type transcends a

diversity in ends offended no one's ontological inhibitions': 'England', in T. F. Glick (ed.), "The comparative reception of Darwinism", Austin & London, 1974, pp.25-6.

69 W. Houghton, "The Victorian frame of mind", new edn., Yale, 1973, p.400.

70 Ibid., pp.400-1.

71 Lyell to Mantell, 8 February 1822; K. M. Lyell, op.cit. (39), i, 115.

72 Cf. Lyell to Herschel, 24 May 1837, ibid., ii, 11.

73 C. Babbage, "Ninth Bridgewater treatise", London, 1837, note D, p.175.

74 Lyell to Herschel, 1 June 1836, in K. M. Lyell, op.cit. (39), i, 467. Whewell added to the ambiguity when he reviewed volume ii of Lyell's _Principles of geology_, for the _Quarterly Review_. New groups of organisms, he wrote, appear 'as if they were placed there, each by an express act of the Creator' (cited by Cannon, 'The problem of miracles', op.cit. (17), p.8). The use of the 'as if' construction did not discourage Lyell himself from hoping that Whewell was close to admitting general laws in this delicate area: see Lyell to Herschel, 24 May 1837, in K. M. Lyell, op.cit. (39), ii, 12, for Lyell's continuing optimism in this respect. The peculiar ability of natural theology to mediate between diverse interpretations of 'natural law' has been studied by J. A. Durant, 'The meaning of evolution: post-Darwinian debates on the significance for man of the theory of evolution, 1858-1908', University of Cambridge PhD thesis, 1977, pp.45-57. For the possibility of fruitful ambiguity in a more squarely scientific context, reference should be made to K. M. Figlio, 'The metaphor of organisation: an historiographical perspective on the bio-medical sciences of the early nineteenth century', Hist. Sci., 1976, xiv, 17-53.

75 Cited by Pye Smith, op.cit. (11), p.428.

76 Ibid.

77 Baden Powell, "Essays on the spirit of the inductive philosophy, the unity of worlds, and the philosophy of creation", London, 1855. Powell was more sympathetic to the theodicy of _Vestiges_ than were most of its critics and he subsequently consolidated his critique of the apologetic value of miracles in his contribution to _Essays and reviews_.

78 H. Gruber, "Darwin on man", London, 1974, p.209.

79 Cited by Pye Smith, op.cit. (11), p.428.

80 Ibid.

81 Ibid., p.429.

82 Ibid., p.428.

83 H. Brougham, "A discourse of natural theology", 3rd edn., London, 1835, p.1.

84 Even Chalmers was induced to comment on the inadequate rewards for British men of science, op.cit. (2), ii, 172.

85 Duke of Argyll, in "Report of the . . . British Association for

the Advancement of Science, 1855", London, 1856, p.lxxxii. For the wider context of the state of scientific education in England and the constraints that still had to be removed in 1870, see M. Gowing, 'Science, technology and education: England in 1870' (The Wilkins Lecture, 1976), Notes Roy. Soc. Lond., 1977, xxxii, 71-90 (85). The Duke of Argyll's remark has been echoed in the judgement of Asa Briggs that 'emphasis by scientists on the social and moral as well as the intellectual or utilitarian role of science was in many ways extremely unwise, for by the time of the revolutions of 1848, it had been demonstrated that "science" might teach the wrong lessons . . . ', op.cit. (48), p.480.

86 J. Playfair, "Illustrations of the Huttonian theory of the earth", Dover facsimile edn., New York, 1964, pp.119-22, 485-6.

87 C. Lyell, "Principles of geology", 3 vols., London, 1830-33, iii, 384-5.

88 Brougham, op.cit. (83), p.199.

89 Ibid., p.204.

90 J. Stephen to Whewell, 15 October and 10 November 1853, Trinity College Cambridge, Add. MSS a 216, 130 & 142.

91 R. Chambers, "Vestiges of the natural history of creation", London, 1844; reprinted Leicester, 1969, pp.331-2, 361-86.

92 Gillispie, op.cit. (17), pp.164-70.

93 It was by a judicious choice of tense that Herschel protected the mediating role of appeals to design when he criticised the thesis of _Vestiges_ in these terms: 'The transition from an inanimate crystal to a globule capable of such endless organic and intellectual development, is _as_ great a step - _as_ unexplained a one - _as_ unintelligible to us - and in any human sense of the word, as _miraculous_ as the immediate creation and introduction upon earth of every species and every individual would be.' "Report of the . . . British Association for the Advancement of Science, 1845", London, 1846, pp. xlii-iii.

94 Cited, with instructive comment, by Yule, op.cit. (20), pp.266-8.

95 A. D. White, "A history of the warfare of science with theology in christendom", 2 vols., New York, 1895; J. W. Draper, "History of the conflict between religion and science", New York, 1875.

96 R. Gruner, 'Science, nature, and Christianity', J. Theol. Stud., 1975, xxvi, 55-81.

97 That this consequence still needs to be explored is evident from the fact that the most recent plea for reevaluation (Bowler, op.cit. (17), pp. 30, 42) has the effect of adding still more to that which Darwin allegedly destroyed.

98 Lyell to Darwin, 11 March 1863, in K. M. Lyell, op.cit. (39), ii, 363. See also Lyell to Hooker, 9 March 1863, ibid., 361. For Lyell's feelings on the descent of man from the brutes, see M. J. Bartholomew, 'Lyell and evolution: an account of Lyell's response to the prospect of an evolutionary ancestry

for man', Brit. J. Hist. Sci., 1973, vi, 261-303.

99 I have developed this point at greater length in Brooke, op. cit. (10).

100 I am simply thinking here of Darwin's reference to the laws impressed on matter by the Creator - a reference prominent in the conclusion of the _Origin of species_. See also Gruber, op.cit. (78), p.209.

101 R. M. Young, 'Natural theology, Victorian periodicals, and the fragmentation of the common context', unpublished MS.

102 Lyell to Fleming, 6 February 1856, in K. M. Lyell, op.cit. (39), ii, 208.

103 The ten were those of Viscount Milton (1832), Buckland as President elect (1832), Sedgwick (1833), Vernon Harcourt (1839), Herschel (1845), Marquis of Northampton (1848), Reverend T. R. Robinson (1849), Brewster (1850), Lord Wrottesley (1860), and Sir William Armstrong (1863).

104 J. D. Hooker, in "Report of the . . . British Association for the Advancement of Science, 1868", London, 1869, p.lxxiv.

105 Ibid.

106 W. Thomson, in "Report of the . . . British Association for the Advancement of Science, 1871", London, 1872, p.cv.

107 In Tyndall's own words: 'we claim, and we shall wrest, from theology the entire domain of cosmological theory.' Reprinted in Basalla, Coleman, & Kargon, op.cit. (14), pp.474-5.

108 A. Thomson, in "Report of the . . . British Association for the Advancement of Science, 1877", London, 1878, p.lxx.

109 Ibid., p.lxxi.

110 Cited in Gordon, op.cit. (5), pp.243-5.

111 Ibid.; see also Briggs, op.cit. (48), p.311.

112 See, for example, the recollections of Augustine Birrell in D. M. Thompson, "Nonconformity in the nineteenth century", London, 1972, p.169; cf. ibid., pp.141-3.

113 K. M. Lyell, op.cit. (39), ii, 134, 86-7, 169.

114 Sedgwick to Miss Gerard, 20 January 1855, in Clark & Hughes, op.cit. (6), ii, 296.

115 Whewell, in "Report of the . . . British Association for the Advancement of Science, 1841", London, 1842, p.xxxiv.

116 Kitson Clark, op.cit. (47), pp.257, 198-205; H. M. Lynd, "England in the Eighteen Eighties", Oxford, 1945, pp.329-43; T. H. S. Escott, "Social transformations of the Victorian age", London, 1897, p.407; Crowther, op.cit. (36), p.10. With particular reference to the proceedings against the authors of _Essays and reviews_, Lyell was able to satisfy himself, in the early 1860s, that churchmen were now 'very free on most points which they were afraid to venture on for fear of legal penalties', K. M. Lyell, op.cit. (39), ii, 360.

117 Hooker to Darwin, 6 October 1865, cited by Gruber, op.cit. (78), p.26.

118 Falconer to his niece, Mrs Grace McCall, 28 April 1863, cited

by P. J. Boylan, 'The controversy of the Moulin-Quignon jaw: the role of Hugh Falconer', this volume.

119 Lyell to Ticknor, 2 April 1847, in K. M. Lyell, op.cit. (39), ii, 127.

120 Ibid., ii, 172. See also p.169 for Lyell's attaching considerable importance to the establishment of a professional system of Divinity at King's College London: 'the first formidable opposition to or competition with the old Universities, and men brought up in the metropolis will have larger views'.

121 Cited by Houghton, op.cit. (69), p.177.

122 K. M. Lyell, op.cit. (39), ii, 455.

123 Huxley to Darwin, 13 January 1862, in L. Huxley, op.cit. (10), i, 281.

124 Houghton, op.cit. (69), p.179; J. S. Mill, "On liberty", 1859; Worlds classics edn., Oxford, 1963, pp.38-43.

125 J. Baldwin Brown, "First principles of ecclesiastical truth", London, 1871, pp.208-364.

126 Ibid., p.227.

127 Ibid., pp.233-5.

128 Ibid., p.244.

129 Ibid.

130 Ibid., pp.278-9.

131 Ibid., pp.279-83.

132 Ibid., pp.320-1.

133 Ibid., p.320.

134 K. M. Lyell, op.cit. (39),ii, 82.

135 Baldwin Brown, op.cit. (125), p.342.

136 Ibid., p.344.

137 Ibid., p.343.

138 Ibid., p.344.

139 Ibid., p.353.

140 To complete the circle it is worth noting that, according to Kingsley, the emphasis on the real Fatherhood of God, which he had learned from Maurice, provided the ground for 'a great deal of my natural philosophy'; Kingsley to Maurice, 1865, in F. E. Kingsley, op.cit. (1), p.267. If it provided only some of the ground for his acceptance of an immanentist version of Darwin's evolution (ibid., p.253), we have striking evidence of the oversimplification which is so commonly built into unidirectional studies of the impact <u>of</u> science <u>on</u> religion.

The language of environmental science

Transposed concepts from the human sciences in the early work of Charles Lyell

MARTIN J. S. RUDWICK

Introduction (1)

The importance of models and metaphors in the construction and articulation of scientific theories should no longer need argument or even emphasis. (2) Historians of science are however becoming more sensitive to the need to explore the sources of such creative analogies. Concepts that are transposed into a field where they have not previously been used, or not used in the same way, are components of cognitive change that must be regarded as 'external', if only to the particular field in question. (3) It is at least arguable that major cognitive innovation is most likely to emerge in the scientific work of individuals who choose to employ analogies that in this sense are strongly 'external': that is, analogies that are furthest removed from the 'normal practice' of the discipline concerned. This may happen when a scientific field scarcely yet deserves the name of 'discipline', because its practice is not yet strongly insulated and institutionalised. It may also happen, however, when the circumstances and personality of an individual incline or encourage him to spread his interests widely, without regard for disciplinary boundaries: he may then be able to see the potential relevance of analogies that would not occur to his more narrowly 'professional' colleagues. To evaluate any particular case of this kind, however, we must survey the whole range of human enterprises which were potentially available to such an individual, and those with which he was actually familiar. We should then try to see how he used these other enterprises as 'cultural resources' from which to quarry the creative analogies that he could shape to fit his own cognitive construction. (4)

In this paper I shall analyse some key analogies of this kind, which recur frequently in the early - and most original - work of one of the central figures of the 'classic' period in the history of geology, namely Charles Lyell. I shall argue that the originality of his conception of what geological science should become was due in part to his creative use of analogies drawn from fields remote from geology. I shall suggest that he found these analogies illuminating for geology precisely because he already appreciated the cognitive power of those other enterprises from which they were drawn. In this way, I shall conclude that Lyell's own cognitive construction must be seen as closely related to his cultural position at the centre of European intellectual life.

The analogy with human historiography

The first analogy I shall explore is the analogy between geology and human historiography. Lyell began his three-volume work the *Principles of geology* (1830-3) with a short introductory chapter, the summary of which began with the words 'Geology defined - Compared to History'. (5) Here he used the analogy with the study of history to outline his belief that geology should be concerned with changes and their causes, and not at all with ultimate origins. He concluded that 'geology differs as widely from cosmogony, as speculations concerning the creation of man differ from history'. (6) The same sentiment can be found in his private notebooks soon after he first began to think about writing a book. But here the phrase 'History of a nation' (7) suggests that the model that Lyell had in mind was a reformed historiography in which grandiose theorising about the origin of mankind was replaced by the more limited goal of a critical reconstruction of the history of particular nations.

Where did Lyell find such a historical model for a reformed geology? In his survey of the history of geology itself, Lyell argued that *no* earlier geological tradition had gained more than partial insights into the true method that the science should follow. (8) He ended this long introduction to the *Principles* by claiming for himself, and for the followers he hoped to attract, that in geology 'the charm of first discovery is our own'. I believe that one reason for this sublime confidence in his own originality can be detected in the rest of that concluding sentence:

> As we explore this magnificent field of inquiry, the sentiment of a great historian of our time may continually be present to our minds, that 'he who calls what has vanished back again into being, enjoys a bliss like that of creating'. (9)

This 'great historian' was Barthold Georg Niebuhr, and Lyell was quoting from the introduction to Niebuhr's _Römische Geschichte._ The rest of Niebuhr's last sentence is, like Lyell's, worth quoting in full (in the translation that Lyell read):

> It were a great thing, if I might be able to scatter, for those who read me, the cloud that lies on this most excellent portion of [hi] story, and to spread a clear light over it; so that the Romans shall stand before their eyes, distinct, intelligible, familiar as contemporaries, with all their institutions and the vicissitudes of their destiny, living and moving. (10)

It would be difficult to improve on this passage from Niebuhr's history, as a summary - in analogical terms - of all that Lyell hoped to achieve in geology: to 'scatter the cloud' of poetic mystery that had been spread over the earth's past, and to throw it into 'clear light'; to 'call what has vanished again into being', recreating the earth's former inhabitants 'living and moving' within their original environments, making them 'distinct, intelligible, familiar as contemporaries'. Niebuhr's history of the Romans provided a perfect model for Lyell's reformed method for recovering the history of the earth.

The first part of Niebuhr's _History of Rome_ was not published in English translation until 1828, by which time Lyell's theoretical position in geology was already established (and he could not have read the original before 1829, when he first learnt German). In general terms, however, he was probably familiar with Niebuhr's work at an earlier date: the newly enthusiastic Germanist Thomas Arnold reviewed Niebuhr's work in the influential _Quarterly Review_ in 1825, only a year before Lyell contributed his own first article to that journal. (11) Anyway Lyell, as a competent classical scholar, would surely have taken an interest in the new German studies of classical antiquity (_Altertumswissenschaft_): at that time it was still novel among English intellectuals, and would have been normal matter for discussion in the circles in which Lyell moved. (12)

In a more limited sense, of course, the analogy with human history was already so well established as to be a commonplace of geological rhetoric. The metaphors of coins and medals, monuments and inscriptions, can be traced back even into the seventeenth century. For our present purpose, however, we need go no further back than Lyell's hero-figure, the French anatomist Georges Cuvier. In the third edition of his great _Recherches sur les ossemens fossiles_, published in 1825 during the most creative period of Lyell's life, Cuvier recalled his earlier pioneer work: 'As a new kind of antiquary, I had to learn . . . to restore these monuments of past revolutions.' (13) Yet 'antiquary' was precisely the right metaphor for Cuvier to use of himself. His anatomical reconstructions of extinct mammals from their fragmentary bones were analogous to an antiquary's reconstruction of (say) a Greek temple, by piecing together the fallen pillars and architraves. Cuvier's work did go somewhat further, since he did try to recreate his extinct animals as living beings that had been adapted to particular modes of life. Yet they were not set within a total reconstructed environment. William Buckland - Cuvier's follower and Lyell's teacher - later extended Cuvier's approach by reconstructing one particular episode more fully. He used impeccable actualistic reasoning to reconstruct the ecological environment and habits of the pre-diluvial cave-hyena - its eating habits, even its defecating habits - with such vivid immediacy that his friends hardly knew where careful inference ended and playful fantasy began. (14) I see no good reason to doubt that Lyell's admiration for Cuvier's palaeontological work was sincere, or that Buckland's vivid reconstruction of the pre-diluvial scene was a model for Lyell's own word-pictures of past environments - this is just where contrasting labels such as 'catastrophist' and 'uniformitarian' can be most misleading. (15) Yet what we find in Lyell's work, far more clearly than in that of any of his elders or contemporaries, is the consistent attempt to turn such isolated 'antiquarian' reconstructions into a truly historical <u>sequence</u> of reconstructions in continual flux.

I have already suggested that the model for this causal and sequential concept of the earth's history came from a deliberate and conscious analogy with human historiography. This is expressed clearly in an essay entitled 'Analogy of geology and history' which Lyell wrote while he was working in the Massif Central in 1828. Here he compared the human history of Auvergne with its earlier geological history of volcanic eruptions: in both cases the observable phenomena failed to make sense unless one recognised 'that the present state of things has grown out of one extremely different'. Lyell believed that to neglect this causal connection of past and present would be to repeat the mistake of earlier historians: 'We err as much as when we judge of a political constitution without considering the pre-existent state of the laws from which it has grown.' (16) Although Lyell did not here refer to him by name, Niebuhr had combatted precisely this error, by tracing in particular the gradual development of the Roman legal system - a point that Lyell, as a lawyer himself, must surely have appreciated. As Thomas Arnold put it in his review article: 'Upon principles such as these Niebuhr has proceeded, and in doing so has adopted the only method by which a real knowledge of Roman history can ever be obtained.' (17) Likewise, I suggest, Lyell realised that the correct 'principles' for geology, the only 'method' that would lead to 'real knowledge' of the earth's history, would be derived from a strictly analogous attention to the sequence of successive states through which the 'system of nature' had passed from epoch to epoch. The causal connection of past and present, in geology as in history, was indeed to be expected; as Lyell expressed it in his essay, 'it belongs to the whole constitution of things, and if natural causes operate as they do, from day to day continually, one epoch must be affected by the events of another'. (18)

This may seem obvious to us or even naive, yet it is important to see how Lyell took this insight from human history seriously, and applied it analogically to geology with a thoroughness that is unmatched among his contemporaries. He was not original in using actualistic reasoning to reconstruct episodes from the past; but only the analogy with human history could underpin his distinctive vision of the earth's past as a continuous causal chain, yielding a dynamic flux of total reconstructed history. This conclusion is not invalidated by the undoubted contrast, on a different level, between Niebuhr's - and Lyell's - 'progressivist' view of human history on the one hand, and Lyell's steady state or 'uniformitarian' model of earth history on the other. For the explanatory goals of both enterprises were deeply causal; and this was not affected by Lyell's adoption - possibly in reaction to the disturbing implications of Lamarck - of a severely steady state geological theory. (19)

The analogy with linguistics

The second analogy I want to mention is the analogy with linguistics. In general terms, of course, the notion of a 'language of nature' is a very old one. Moreover, like the metaphors of the antiquary and his 'monuments', it was also well established by Lyell's day in the more specific sense of a means of deciphering the *past* history of nature. For example, when Cuvier termed himself a new kind of antiquary, he described his task as both to restore these monuments of past revolutions, and '*to decipher the meaning of them*'. (20) In other words, the 'coins' and 'monuments' of nature could yield little insight unless the 'inscriptions' on them could be 'deciphered' and 'read'.

Lyell adopted this metaphor enthusiastically. For example, shortly after his essay on the history of Auvergne, he wrote as follows about his work with Roderick Murchison near Vicenza: 'The volcanic phenomena were just Auvergne over again, and we read them off, as things written in a familiar language, though they would have been Hebrew to us both six months before.' (21) Lyell's adoption of the metaphor of linguistic decipherment is important, because it suggests how he was able to escape from the stultifying empiricism and aversion to theorising that characterised his main scientific milieu, the Geological Society of London. For the metaphor implied a recognition of the essentially interpretative or hermeneutic task of the geologist. On this view, 'Baconian' fact collecting was impossible in geology, or anyway useless, for the 'documents' of the past history of the earth had to be 'read' in a 'language' that had to be learnt.

Another example may illustrate this point more clearly. Lyell's work in Italy in the winter of 1828-9 convinced him that fossil molluscs held the key to an understanding of the relatively recent 'Tertiary' epochs of the earth's history, and perhaps of earlier epochs too. On his way back to England, he therefore stayed in Paris to study conchology under the guidance of Paul Gerard Deshayes. Writing home about his work with fossil shells, he said: 'It is the ordinary, or as Champollion says, the demotic character in which Nature has been pleased to write all her most curious documents.' (22) Here the immediate reference was to the newly successful decoding of Egyptian hieroglyphs, using the demotic script of the Rosetta stone as a linguistic 'bridge' to the Greek text. But Lyell's use of this linguistic analogy emphasises how he saw fossil molluscs as a natural 'language' which had to be deliberately learnt; at the same time, perhaps, he implied that this language was easier to decipher, and a more reliable guide, than the more problematic 'language' of the spectacular fossil vertebrates that Cuvier had reconstructed.

Later, when writing the Principles, Lyell repeatedly used the metaphors of the 'documents' or 'inscriptions' on nature's 'monuments' as a means of illustrating the fragmentary character of the 'records' or 'annals' that nature had left. He stressed the need to learn the full vocabulary of the 'living language' of nature, or in other words the full extent of geological 'causes now in operation', before the records of past events could be correctly interpreted. (23) Even Cuvier himself, Lyell suggested, had been inconsistent in concluding prematurely that 'the thread of induction was broken' between present and past, for Cuvier's successful reconstructions of extinct animals had been 'an acknowledgement, as it were, that a considerable part of the ancient memorials of nature were written in a living language'. (24)

At one point, Lyell applied the linguistic analogy in a significantly different way. Like many of his contemporaries, he was deeply struck by the rediscovery of Herculaneum under the foundations of Resina in the shadow of Vesuvius. He imagined an antiquary of a future period discovering a still fuller sequence of three buried cities, one above the other, in which inscriptions might show that the language had been Greek in the oldest city, Latin in the second and Italian in the youngest - a reasonable possibility in Campania. Lyell used this thought-experiment to suggest that:

> the catastrophes, whereby the cities were inhumed, might have no relation whatever to the fluctuations of the language; and that . . . the passage from the Greek to the Italian may have been

very gradual, some terms growing obsolete, while others were introduced from time to time. (25)

The primary purpose of this illustration was to emphasise by analogy the fragmentary nature of the geological record, and the need to distinguish occasional accidents of preservation from the original sequence of continuous processes. But there are two incidental features about Lyell's thought-experiment that are worth noting.

First, it shows even more clearly than my other examples that Lyell was aware of human languages as changing dynamic systems. Like his awareness of German historiography, his familiarity with recent developments in linguistics needs no special explanation: it was a subject of general interest in the intellectual circles in which he moved, and Lyell himself as a classical scholar would naturally have shared this interest. Certainly he was familiar enough with Sir William Jones's pioneer research on Sanskrit to use Jones's edition of the _Institutes of Hindoo law_ as the first item in his review of the history of geology in the _Principles_. (26) He may also have known of the more recent German philology of Grimm, Bopp and others, which became well known in England after Rask published a review of Grimm's work in 1830. (27)

The idea of language as a continually changing system was thus available to Lyell as a conceptual resource from his general intellectual environment. Yet it is significant - and this is my second point - that he applied it only in a limited way. He does not seem to have grasped one essential feature of the new linguistics, namely the idea that words themselves have undergone imperceptibly gradual changes in the course of time. In his geological analogy, Lyell seems to have taken words as the basic units of language, units which could be 'introduced' or 'grow obsolete' but which did not themselves change. The language as a whole could change very gradually, but only by piecemeal change in the constituent units, the words. This model is strikingly parallel to Lyell's image of piecemeal change in faunas and floras, whereby the totality of organisms could change very gradually, but only by the introduction and extinction of the constituent units, the species. This parallel suggests that Lyell's insistence on the reality of unchanging units in biology may have had other reasons besides a desire to avoid or forestall evolutionary explanations. It may have been rooted more deeply in the structure of his thinking, in the sense that he may have tended to see _all_ natural phenomena in terms of clearly defined natural kinds; he may have been unable to conceive any blurring of these basic categories, even when such 'transmutations' would have had minimal implications for his view of man in nature. (28)

The analogy with demography

The third analogy is that with demography or populational analyses. One essential component of Lyell's use of this in his geology was his concept of a biological species as an analogue of an individual organism. The initial value of this analogy was that it suggested a possible explanation of the phenomenon of extinction, the reality of which Cuvier's research had demonstrated for the first time beyond reasonable doubt. If species, like individuals, had intrinsically limited life-spans, then extinction might be explained without recourse to mere chance or accident. Lyell was using this analogy even in his first published essay on geology in 1826. Here he suggested that extinction was 'a phenomenon perhaps not more unaccountable than one with which we are familiar, that successive generations of living species perish, some after a brief existence of a few hours, others after a protracted life of many centuries'. (29) Lyell's phrasing here is so close to that used by Giovanni Battista Brocchi some ten years previously that it seems almost certain he had already read Brocchi's 'Reflections on the extinction of species'. (30) This is not a matter of trivial source-hunting: it is important because it implies that this crucial analogy between species and individuals was available to Lyell at an early date in a context that linked it explicitly with the problem of Tertiary geology, for Brocchi's essay on extinction concluded the introduction to his great _Conchiologia fossile subappenina_ (1814). Certainly Lyell was applying the analogy by the time he was working in the Massif Central in 1828, for in his essay on the history of Auvergne he mentioned the difficulty of determining 'the laws which regulate the comparative longevity of species'; (31) and later, in the _Principles_, he wrote quite casually about the 'birth and death of species'. (32)

I cannot here go into the highly important and much debated question of Lyell's early views on the origin of species. But it is worth noting in passing that the analogy between species and individuals did give him an outline model for organic change that would be naturalistic and yet non-Lamarckian. If species were in some sense 'born', they must have arisen by natural means from some pre-existing species; yet the analogy also implied that species might appear at definite points in time, with an individuality that would distinguish them from other species and characterise them throughout their life-span. It should go without saying that this naturalistic model would have been perfectly compatible in Lyell's mind with a belief that such events must have been supervised providentially by an omniscient Creator. (33)

For our present purpose, however, what is more important is the way that Lyell extended this essentially biological analogy between species and individuals into a more complex model of organic change.

In doing so, he made significant use of an analogy with demography. The stages by which he developed this analogy are still obscure; but in the last volume of the Principles it formed the core of his interpretation of the scattered Tertiary strata of Western Europe, which in turn was presented as a model example of his method of reconstructing the past history of the earth as a whole.

Lyell introduced the analogy simply as an illustrative device: 'Let the mortality of the population of a large country represent the successive extinction of species, and the births of new individuals the introduction of new species.' He then imagined itinerant 'commissioners' visiting the different provinces of the country in succession, taking a series of censuses and then leaving behind them a series of 'statistical documents' for each province. The recorded changes in the population of a given province would then be proportional to the lapse of time since the previous census, subject to disturbing factors such as epidemics or immigration. This, Lyell argued, would be analogous to the observed changes in the assemblages of species preserved in the various 'basins' of Tertiary strata. (34)

This was an effective analogy for Lyell to use: regular decennial censuses had been introduced in Britain in 1801, so that by the time he was writing there had already been a succession of four such periodic censuses. (35) The system had been introduced precisely in order to discover whether the population was changing in size, and if so, how, where, and in what direction. The censuses were the empirical counterpart to the famous Malthusian debates on social policy in relation to population. Once again there is no special problem about Lyell's familiarity with these questions. Although he was too much a dedicated scientist to get involved in politics himself, he could hardly avoid being aware of political issues of such central importance. (36) Furthermore - to mention just one example - one of his closest geological friends and allies, George Poulett Scrope, was also a prominent political economist in the anti-Malthusian camp.

I suggest, therefore, that Lyell used the current Malthusian debate as a readily available source for an analogy that was of central importance in his geology: namely, the demographic model of a human population in continual flux, as an analogue of the continual changes in the earth's fauna and flora through the epochs of geological time. In view of the recent controversy about the ideological implications of Darwin's later use of Malthus, (37) it is worth pointing out that Lyell's use of a demographic model was neutral with respect to any concept of struggle or disharmony in nature or society. This does not imply the Lyell himself had no opinion on these matters: only that his position vis-à-vis Malthus had no discernible effect on his use of a demographic analogy in geology.

The analogy with political economy

This brings me to the closely related analogy with political economy. I have suggested elsewhere how a casual metaphor in Scrope's geological work - the idea that geologists must make 'almost unlimited drafts upon antiquity' - shows that he was using a parallel between money and time as an analogy of great heuristic power; but there is no evidence yet that the analogy operated in this form outside Scrope's own mind. I also suggested, however, that in a more general way the statistical approach of political economy, as developed in the work of Scrope and his contemporaries, may have been profoundly influential on the geology of Lyell. (38) On both sides of the Malthusian debate, political economists saw the whole economic system as one of antagonistic forces, operating through a complex mass of individual decisions and events. How far Lyell was familiar with theoretical political economy I do not know, though he was reading Adam Smith's *Wealth of nations* as early as 1823. (39) Yet in any case the political economists' image of a complex balanced system is strikingly parallel to Lyell's overall vision of the geological system. It is probable, of course, that Lyell derived the general concept of balanced antagonistic forces in geology from the natural philosophy of Adam Smith's Edinburgh friend James Hutton, as mediated through John Playfair's *Illustrations of the Huttonian theory of the earth* (1802). Yet that orthodox interpretation of intellectual influences fails to acknowledge one subtle but profound contrast between the Huttonian tradition and Lyell's work. Like the 'conjectural history' of the Scottish Enlightenment, Hutton's system - and Playfair's re-interpretation of it - was one of 'conjectural' or idealised continents and oceans. It was almost as an afterthought, for example, that Hutton mentioned that the processes of erosion, sedimentation and elevation would not really be successive but simultaneous. (40) Lyell, on the other hand, sought to reveal a more varied battery of geological forces by means of a highly concrete analysis of changes observable in the present or the recent 'historic' past. The units of his analysis were never 'conjectural' continents and oceans, but always specific and particular volcanoes,river valleys, earthquakes, ocean currents and so on. In his application of 'causes now in operation' to the explanation of the more remote pre-human past, he left his readers in no doubt that he believed that the dynamic equilibrium of the earth was the result of a summation of innumerable concrete local events of the same kind. It is this that gives Lyell's system a stochastic character that places it much nearer the political economy of his contemporaries like Scrope, than to the earlier natural philosophy of Hutton.

At the same time, however, a comparison between Lyell's work and that of Karl Ernst Adolph von Hoff shows how the theoretical framework of balanced antagonistic forces - whether derived from the

Huttonians or not - was as essential to Lyell's system as concrete local data on geological processes and their results. Alexander von Humboldt recommended Lyell to read von Hoff, and Lyell learnt German specifically in order to do so. (41) He used von Hoff's work extensively; but when Scrope was about to review the Principles, Lyell told him: 'You should compliment him [von Hoff] for the German plodding perseverance with which he filled two volumes with facts like statistics, but he helped me not to my scientific view of causes.' (42) This was somewhat bluntly expressed, but nevertheless strictly accurate. Von Hoff listed his data under headings such as 'extension of the area of the sea' and 'extension of the surface of the land', without distinguishing between the many different processes that had produced such extensions of the sea or land during historic times. Thus although Lyell could use von Hoff's work profitably as a quarry for examples, its arrangement totally obscured the dynamic interplay of antagonistic forces which Lyell wished to expound.

It is significant that Lyell described von Hoff's work as containing 'facts like statistics'. In the original sense of the word, 'statistics' were the empirical data on which it was hoped that the 'statist' or statesman could base rational policies. The fact-collecting model of 'statistics' would therefore have come naturally to a professional diplomat and civil servant like von Hoff. Yet it is important to note how far Lyell transcended von Hoff's work. Although he appreciated the value of such factual documentation in geology, Lyell saw the limitations of the empiricism of contemporary 'statistics', and he used von Hoff's data in the service of a far more theoretical enterprise. I therefore suggest that what we can see in Lyell's work, in this respect, is a creative fusion between the abstract 'conjectural' model of a natural system in dynamic equilibrium, and the stochastic perspective that contemporary political economy derived from empirical 'statistics'.

Conclusion

In this paper I have outlined Lyell's use of four major analogies drawn from outside geology, and indeed from outside the natural sciences altogether. At the very least, these analogies were deployed effectively in his published work as a means of communicating his conceptual vision to his readers. But the distinction between analogical communication and analogical thinking is blurred if not unreal, for a writer can scarcely employ an analogy with conviction if it has not already proved its worth in his own thinking. It is therefore reasonable to conclude that these analogies functioned not only as persuasive images for his readers but also as heuristic images for Lyell himself. Historiography strengthened his awareness of the deep causal connectedness of past, present and future, and the need for critical scrutiny of fragmentary evidence. Linguistics strengthened his awareness that geology must be interpretative or hermeneutic throughout, and that the 'languages' of nature needed to be learnt. Demography strengthened his awareness that gradual overall change could result from the summation of piecemeal changes in innumerable discrete units. Political economy strengthened his awareness that innumerable local events of a stochastic character could add up to a total system that was providentially in harmonious balance.

Lyell's familiarity with these diverse fields of contemporary study in the human sciences was simply a result of his own broad interests. These in turn were related to his favourable milieu at the centre of English metropolitan intellectual life. In this respect his membership of the Athenaeum in London - he joined the club in 1824 soon after its foundation - may well have been as important as his membership of the Geological Society. (43) Indeed, since he was a good linguist and also financially in a position to travel extensively all over Europe, he was in personal contact with *savants* in Paris and other cultural centres, so that we can say that his milieu was at the centre of European intellectual life.

Lyell's cultural milieu made accessible to him the resources from which he shaped the creative analogies I have discussed here; but his milieu is in itself no adequate explanation of his use of these analogies. After all, most of the other individuals who dominated British geology around 1830 - men such as Murchison, Sedgwick and Buckland - were also gentlemen of liberal education and broad interests. Yet none of them produced a synthesising system of geology as comprehensive and coherent as Lyell's, nor did they employ a wide range of analogies drawn from outside the natural sciences. A possible explanation of this is that Lyell was simply less 'professional'. Since the concept of 'professionalism' in the history of science is too easily equated with scientific excellence, this suggestion needs to be amplified. Lyell's Parisian and German-speaking contemporaries were, by and large, 'professional' geologists in the strict sense that they were paid to practice geology in one way or another for at least part of their time. Fewer of Lyell's major colleagues in England were 'professionals' in this sense; but the group that dominated the Geological Society in London did exercise a very real control over the standards and procedures of their science. It would be anachronistic to term them scientific specialists; but they did have a clear conception of a defined field of 'geology', the credibility of which could only be guaranteed by keeping a close watch on what was done within its boundaries. With this implicit concept of intellectual guardianship, I suggest that they felt neither need nor inclination to import analogical resources across the cognitive frontier of geology, even if they were familiar - as they were - with contemporary studies such as those that Lyell used.

Why then did Lyell do what his contemporaries failed to do, and make creative use of heuristic and persuasive analogies drawn from outside geology? He had as much concern as any of his contemporaries for establishing geology on firm foundations. Yet I suggest that he saw his geology as a much more open-ended enterprise. His pursuit of geology was the focal point of his chosen way of life, but he seems to have had little sense of cognitive compartments and boundaries. In his published work he always presented himself - perhaps deliberately - as the broadly-educated man of culture, not the narrow specialist in geology. (44) Even within the natural sciences, he quarried his evidential materials from far wider sources than most of his contemporaries - from physical geography, meteorology, many branches of biology, even the physical sciences (where he perhaps felt least at home) - without any sense that he was improperly trespassing into foreign territory. So his reaching into realms beyond the natural sciences altogether was only a further extension of the same 'uncompartmentalised' attitude to human knowledge as a whole.

This free-ranging attitude cannot simply be attributed to his financial position. He was indeed sufficiently independent not to be constrained by any narrowly 'professional' concern with building a career; but his most significant fellow-geologists, at least those in England if not the French, shared that independent position. But Lyell seems to have been far less concerned with constructing an autonomous new science than most of his contemporaries. He did indeed want to reformulate the basic approach to be used by geologists, but he saw this as part of a much wider cognitive enterprise. In this sense, there was indeed an affinity with the earlier generation of Huttonians: Lyell was still more of a natural philosopher than a geologist. With this broad outlook, he not only felt an interest in a wide range of other cognitive enterprises, including many fields outside nature science altogether; he also appreciated their cognitive power, and saw geology as a field that could learn from their success. Hence, unlike his more 'professional' colleagues, Lyell was able to quarry creative analogies from a wide range of extra-geological sources, and to shape them effectively to fit his own construction of a reformed approach to geology.

In this respect, Lyell's work forms an instructive example of how a 'boundary-ignoring' scientist, by the very breadth of his intellectual concerns, can influence the methodology and cognitive content of even a relatively well established 'discipline' in natural science.

Notes

1 An earlier version of this article was given as a lecture during a symposium in honour of R. Hooykaas in Utrecht in March 1977, and published as 'Historical analogies in the early geological work of Charles Lyell', Janus, 1977, lxiv, 89-107. In revising the article, I benefited greatly from the discussion of the original version during the conference on 'New perspectives in the history of geology' in Cambridge in April 1977. The central sections of the article have been revised in relatively minor ways, but the interpretative context has been radically changed by a completely new introduction and conclusion.

2 In the extensive philosophical literature on this theme, I find the work of Mary B. Hesse especially helpful in a historical context: see her "Models and analogies in science", London, 1963, and "The structure of scientific inference", London, 1974, especially 'Theory as analogy', pp.197-222.

3 On this process of transposition, see for example the already classic work of Donald A. Schon, "Displacement of concepts", London, 1963; republished as "Invention and the evolution of ideas", London, 1967.

4 For an outstanding case study in the history of science, using this 'constructivist' perspective, see H. E. Gruber & P. H. Barrett, "Darwin on man: a psychological study of scientific creativity", New York & London, 1974, especially Book I, Part II, 'The development of Darwin's evolutionary thinking'.

5 C. Lyell, "Principles of geology, being an attempt to explain the former changes of the earth's surface, by reference to causes now in operation", London, 1830-3, i, 1. I use the first edition of each volume throughout this article.

6 Lyell, op.cit. (5), i, 4.

7 L. G. Wilson, "Charles Lyell, the years to 1841, the revolution in geology", New Haven, 1972, p.172; quoted from a notebook dated about July 1827.

8 R. Porter, 'Charles Lyell and the principles of the history of geology', Brit. J. Hist. Sci., 1976, ix, 91-103. I now agree that Lyell was not claiming intellectual ancestry even among the Huttonians; compare M. J. S. Rudwick, 'The strategy of Lyell's Principles of Geology', Isis, 1970, lxi, 5-33 (8-9).

9 Lyell, op.cit. (5), i, 74.

10 B. G. Niebuhr, "The history of Rome", (tr. J. C. Hare & C. Thirlwall), vol.i, Cambridge, 1828; (original edition, "Römische Geschichte", Berlin, 1811-12).

11 Anon. [T. Arnold], 'Römische Geschichte von B. G. Niebuhr . . .', Quart. Rev., 1825, xxxii, 67-92.

12 See for example D. Forbes, "The Liberal Anglican idea of history", Cambridge, 1952, chapter III; K. Dockhorn, "Der deutsche Historismus in England", Göttingen, 1950.

13 G. Cuvier, "Recherches sur les ossemens fossiles", 3rd edn., Paris, 1825, vol.i, 'Discours sur les révolutions du globe', p.l. Compare G. Buffon, "Époques de la nature", Paris, 1778, 'Premier discours', p.l.

14 W. Buckland, "Reliquiae diluvianae", London, 1823. See for example Lyell's reaction, quoted in Wilson, op.cit. (7), p.95.

15 I hope to discuss elsewhere a possibly more fruitful way of 'mapping' these and similar theoretical positions, so as to bring out the substantial similarities between (for example) Lyell on the one hand and Cuvier and Buckland on the other, while preserving their scarcely deniable contrasts.

16 Extracts from Lyell's essay are printed in Wilson, op.cit. (7), pp.215-16.

17 Arnold, op.cit. (11), pp.71-2. Porter, op.cit. (8), points out that Lyell conspicuously failed to apply Niebuhrian standards to his historical account of geology itself: here a 'catastrophist' historiography suited his polemical purposes better!

18 See (16).

19 For the suggestion that Lyell shifted from a 'directionalist' to a 'steady state' position in geological theory as a direct result of realising the implications of Lamarck's evolutionary theory for the place of man in nature, see M. Bartholomew, 'Lyell and evolution: an account of Lyell's response to the prospect of an evolutionary ancestry for man', Brit. J. Hist. Sci., 1973, vi, 261-303.

20 Cuvier, op.cit. (13), my italics.

21 K. M. Lyell, "Life, letters and journals of Sir Charles Lyell, Bart.", 2 vols., London, 1881, i, 203.

22 K. M. Lyell, op.cit. (21), i, 251.

23 Lyell, op.cit. (5), iii, 461-2.

24 Ibid., i, 72-3.

25 Ibid., iii, 33-4.

26 Ibid.,i, 5-6.

27 See H. Aarsleff, "The study of language in England, 1780-1860", Princeton, 1967, chapters IV, V; J. Burrow, 'The uses of philology in Victorian England', in R. Robson (ed.), "Ideas and institutions of Victorian Britain", London, 1967, pp.180-204.

28 My suggestion alludes indirectly to the quasi-structuralist position of M. Foucault's "Les mots et les choses. Une archéologie des sciences humaines", Paris, 1966; (tr., "The order of things. An archaeology of the human sciences", London, 1970). But I would interpret the 'structure' of Lyell's thinking as intermediate between a more static or 'grid'-like mode of knowledge and a fully dynamic or 'flux'-like mode. Lyell (and any other individual thinker) must be located historically, and not assigened tout court to one side or the other of a catastrophe-like 'epistemic break'. Any adequate historical account of broad epistemic changes must allow for the piecemeal character of such changes and for the great diversity of thinking

and knowing at any one period. To deny this, as Foucault's work often seems to, is to abandon any hope of relating cognitive structures to their social environment.

29 Anon. [C. Lyell], 'Transactions of the Geological Society of London . . . ', Quart. Rev., 1826, xxxiv, 507-40 (538).

30 G. B. Brocchi, "Conchiologia fossile subappenina", Milano, 1814: Brocchi discussed the 'unequal span allotted to individuals of different species: the may-fly lives only a few hours, while the stag can maintain itself for several centuries' (i, 227-8). It is curious to find, at this late date, a reference to the legendary long life of 'il cervo', but the general meaning is clear. For Lyell's similar 'borrowing' from Brocchi, this time for his history of geology, see P. J. McCartney, 'Charles Lyell and G. B. Brocchi: a study in comparative historiography', Brit. J. Hist. Sci., 1976, ix, 175-89.

31 Quoted in Wilson, op.cit. (7), p.215. The quoted phrase is not the title of a separate essay, but comes in the middle of a sentence in Lyell's essay on 'Analogy of geology and history'. I am greatly indebted to the Lady Lyell for supplying me with a copy of the relevant pages of Lyell's notebook, which enabled me to check this point.

32 Lyell, op.cit. (5), iii, 32-3. In ibid., ii, 37, Lyell used the same analogy to explain his concept of intra-specific variation.

33 For an unusually clear expression of Lyell's providentialist view of the origin of species, see M. Rudwick, 'Charles Lyell speaks in the lecture theatre', Brit. J. Hist. Sci., 1976, ix, 147-55 (149-50). The Anglican context of his lecture may explain why Lyell chose to express his view so clearly on this occasion, but I see no adequate grounds for dismissing it as insincere: see M. Rudwick, 'Charles Lyell, F.R.S. (1797-1875) and his London lectures on geology, 1832-33', Notes Roy. Soc. Lond., 1975, xxix, 231-63.

34 Lyell, op.cit. (5), iii, 31-3. For the context of the analogy, see Rudwick, op.cit. (8), and 'Charles Lyell's dream of a statistical palaeontology', Palaeontology, 1978, xxi, 225-44.

35 D. V. Glass, 'Some aspects of the development of demography', J. Roy. Soc. Arts, 1956, civ, 854-69; M. J. Cullen, "The statistical movement in early Victorian Britain", Hassocks, England, 1975, pp.10-13.

36 For Lyell's attitude, see J. B. Morrell, 'London institutions and Lyell's career, 1820-41', Brit. J. Hist. Sci., 1976, ix, 132-46, and Rudwick, 'Lyell and his London lectures', op.cit. (33).

37 See for example R. M. Young, 'Malthus and the evolutionists: the common context of biological and social theory', Past and Present, 1969, xliii, 109-45; and 'The historiographic and ideological contexts of the nineteenth-century debate on man's place in nature', in M. Teich and R. M. Young (eds.), "Changing perspectives in the history of science", London, 1973, pp.344-438.

Compare P. J. Bowler, 'Malthus, Darwin and the concept of struggle', J. Hist. Ideas, 1976, xxxvii, 631-50.

38 M. J. S. Rudwick, 'Poulett Scrope on the volcanoes of Auvergne: Lyellian time and political economy', Brit. J. Hist. Sci., 1974, vii, 205-42 (236-42).

39 Wilson, op.cit. (7), p.115.

40 J. Hutton, "Theory of the earth", 2 vols., Edinburgh, 1795, i, 195-7. Significantly, Lyell was among the many readers of Hutton (and/or Playfair) who overlooked this disclaimer: see Lyell, op.cit. (5), i, 473.

41 Wilson, op.cit. (7), pp.170, 276.

42 K. M. Lyell, op.cit. (21), i, 268-9. Only the first two volumes (1822-4) of von Hoff's "Geschichte der durch Ueberlieferung nachgewiesenen natürlichen Veränderungen der Erdoberfläche", Gotha, 1822-41, were published before the first edition of the Principles.

43 Wilson, op.cit. (7), p.35, gives the date when he joined the Athenaeum.

44 See Porter, op.cit. (8).

Geology and landscape painting in nineteenth-century England

MARCIA POINTON

> Does then the artist concern himself with microscopy, History, Palaeontology?
>
> Only for purposes of comparison, only in the exercise of his mobility of mind. And not to provide a scientific check on the truth of nature.
>
> Only in the sense of freedom. (1)

The objects of this essay are twofold. In the first place,we are concerned to define the relationship between British landscape painting and geological theories and discoveries relating to the earth's history that were widely disseminated in the middle years of the nineteenth century. In the second place, the use that could be made of this knowledge by exploiting it in terms of visual imagery will be demonstrated through a detailed critical analysis of paintings by two educated and materially successful Victorian artists. The tension between representation and creation, between mimesis and invention which Klee postulated in the first quarter of our own century can be seen as the pivot on which the nineteenth-century debate balances, whether its mode be verbal or pictorial. We are, in short, dealing with the dilemma of an age when the artist is exposed to novel views about the nature of his environment but when, at the same time, representational and objectively viewed landscape is regarded as a low form of art unsuited to the expression of universal truths and ideas. In order to understand this dilemma we must first look briefly at the more recent history of landscape painting.

In 1771, Sir Joshua Reynolds, speaking to the Royal Academy, said of the seventeenth-century French painter Claude:

> Claude Lorrain . . . was convinced, that taking nature as he found it seldom produced beauty. His pictures are a composition of the various draughts which he had previously made from various beautiful scenes and prospects . . . That the practise of Claude Lorrain, in respect to his choice, is to be adopted by landscape painters in opposition to that of the Flemish and Dutch schools, there can be no doubt, as its truth is founded upon the same principle as that by which the Historical painter acquires perfect form. But whether landscape painting has a right to aspire so far as to reject what the painters call Accidents of Nature, is not easy to determine. It is certain Claude Lorrain seldom, if ever, availed himself of these accidents; either he thought that such were contrary to that style of general nature which he professed, or that it would catch the attention too strongly, and destroy that quietness and repose which he thought necessary to that kind of painting. (2)

Reynolds's main concern in his discourses is with History painting - that is with morally elevating subjects from literature and history treated in the grand style of Raphael and Michelangelo. Landscape was a lowly genre of art but, in so far as it was admitted in the canon of taste, Reynolds's view of the relative merits of Claude and painters of the Dutch and Flemish schools remained virtually law until the end of the eighteenth century and was widely influential well into the nineteenth. We are referring here to British art but in fact similar academic tenets were strong in most European countries at this time.

In addition to the choice between Claudian tranquillity or the rude cottages and blasted trees favoured by, for example, Jacob van Ruisdael and Wynants, there were two further possible landscape modes available to the student of art in Reynolds's time. Salvator Rosa was widely admired for the dramatic way in which he exploited those accidents of nature which an Aristotelian approach to landscape excluded and his work became, not surprisingly, extremely popular with the authors and readers of Gothic novels towards the end of the eighteenth century. (3)

The other possibility open to the landscape artist was never regarded as anything more than hack work by writers on art from Reynolds to Ruskin. Nevertheless, topographical landscape drawing and painting was an important industry throughout the seventeenth and eighteenth centuries. Attention has been drawn to the ability of eighteenth-century geologists to draw field sketches with a degree of competence as a factor of some importance in the development of the science. (4) Certainly it is true that the leisured social classes from which many of these early geologists came were, through their education and environment, equipped in draughtsmanship. But the menial status of the topographer needs to be emphasised in this discussion. The publication and wide dissemination of books on picturesque travel in the 1780s was a peculiarly British phenomenon which had a very great influence on artists, writers and the public alike. (5) Landscape painting meant works by or in the manner of Claude Lorrain, Salvator Rosa, Rubens, or Rembrandt, and any attempt to render the specific truths of nature in particular, as opposed to the divine artistry of Mother Nature was despised as 'mapwork'. Professor of Painting at the Royal Academy 1799-1805 and 1810-25, Fuseli summed up the attitudes of his age when he described 'as the last branch of uninteresting subjects, that kind of landscape which is entirely occupied with the tame delineation of a given spot . . .'(6)

Since the accurate recording of features of the landscape without improvement or embroidery was essential to the geologist (who in some cases undertook these topographical studies himself and in other cases took artists along with him), one might reasonably expect the empirical tradition of the topographer to have had the greatest influence on the development of landscape painting in the nineteenth century, the period when geology becomes a science of major importance.

The tradition of topography was one in which the main aim of the artist was to reproduce accurately what he saw without embellishment and without the sort of refinement and analysis which artists like Girtin and Turner introduced into watercolour painting at the end of the eighteenth century as they felt their way towards a visual language capable of expressing not only the features of the landscape but also their personal response to the changing facets of the scene. But topographical artists, whose main tasks had been antiquarian or military (the recording of ancient buildings, harbours and coastlines) used an outline technique which was not well suited to the needs of the geologists. In fact, it is to those undisciplined artists who observed and incorporated into their pictures the accidents of nature Claude was thought to have rejected that we must first turn.

In order to convey changes in the level and in the geological character of a terrain it was necessary to employ a more painterly technique than topographical artists had used. Gilpin advocated roughness and brokenness which led artists in pursuit of the picturesque to pay greater attention to texture in their landscapes and to abandon large unbroken colour washes. However, it was the coupling of Burke's theories of the sublime to landscape that really opened the way to a union between art and geology. (7) It was not a question, as Rudwick suggests, of the accurate depiction of geological formations (in, for example, views of the Isle of Staffa and of The Needles, from the second decade of the nineteenth century) as opposed to a picturesque view. (8) Actually, eighteenth-century writers on science and those like Gilpin and Alexander Cozens who were preoccupied with theories of the picturesque shared a concern with classification and analysis. (9)

Topography was concerned with accuracy and this was important to mid-nineteenth century landscapists, but the tradition of topographical draughtsmanship did not provide an adequate visual language for a geological age. This had to come from the picturesque and, above all, from the sublime. The convincing evocation of scale is, it has been pointed out, the first requisite for the representation of the sublime. (10) Scale requires space and both are vital to any representation that pretends to convey either the scientific or the emotional truth of a geologically observed landscape. Moreover, the theories of the sublime that Burke had disseminated so widely were based on the principle of association. The sublime consists of an effect upon the human imagination and certain categories are more likely than others to produce these effects. One of these categories can most simply be described as accidents of nature, notable geological features like stupendous cliffs and overhanging precipices. (In this context Salvator Rosa becomes the model rather than Claude Lorrain.) Thus, on one level, the growing interest in geology in the nineteenth century was readily absorbed into an existing tradition remote from topography and the ground was prepared for an alliance between landscape painting and geology which would operate as much through the imagination as through empiricism.

Travel was as vital to the artist as to the geologist but there was always an overlap between documentation and creation. When the artist William Hodges accompanied Captain Cook on his second voyage round the world, his task was undoubtedly topographical in intention (that is, it was to transcribe literally the features he saw), but the file of tiny ant-like human beings struggling up the side of the inner crater transforms this bird's eye view of Mauna Loa in Hawaii into a view that might reasonably qualify as sublime. (11) James Ward's _Gordale Scar_ (1813), (12) generally regarded as the prototype of the romantic landscape, is a direct attempt to transcribe Burke's theories into paint but, in the process, Ward also represents, with considerable accuracy, a specific and well known feature of geological interest. _Gordale Scar_ represents the synthesis of a new 'scientific' attitude to observation with a high degree of intensely subjective feeling. This combination is important for the development of landscape painting throughout the nineteenth century and is expressed most memorably in Ruskin's writing and in paintings of the fifties and sixties.

Ruskin's father commented that his son had been an artist from boyhood but a geologist from infancy. Ruskin's own drawings demonstrate a remarkable degree of sensitivity and skill. His writing, however, in the first place manifests a deep-rooted belief in the aesthetics of the wonderful and the rhetoric of the mysterious that had been the basis of landscape depiction, literary and pictorial, since the second half of the eighteenth century, and in the second place stresses the moral end of landscape painting, thus reminding us not only of the sublime but also of academic theory expressed by Reynolds:

> Whatever influence we may be disposed to admit in the great works of sacred art, no doubt can, I think, be reasonably entertained as to the utter inutility of all that has been hitherto accomplished by the painters of landscape. No moral end has been answered, no permanent good effected, by any of their works. They may have amused the intellect, or exercised the ingenuity, but they have never spoken to the heart. Landscape art has never taught us one deep or holy lesson; it has not recorded that which is fleeting, nor penetrated that which was hidden, nor interpreted that which was obscure; it has never made us feel the wonder, nor the power, nor the glory of the universe; it has not prompted to devotion, nor touched with awe; its power to move and exalt the heart has been fatally abused . . .
>
> It is not, therefore, detail sought for its own sake - not the calculable bricks of the Dutch house painters, nor the numbered hairs and mapped wrinkles of Denner, which constitute great art, - they are the lowest and most contemptible art; but it is detail referred to a great end, - sought for the sake of inestimable beauty which exists in the slightest and least of God's works . . . (13).

Long after Lyell's explanatory work was known, the aesthetics of the sublime continued to be the basis on which people viewed geological features as mysterious, awe-inspiring, theatrical, and wonderful:

> Geology has scattered over plains many a 'wonder' for generations unborn; such as the scene represented which a correspondent at Sidney (who has favoured us with a sketch) describes as 'one of the most remarkable and mysterious features in Geology ever yet discovered'. The moon has lent her mystic influence to the scene. (14)

In attempting to establish the way in which geological knowledge was absorbed into an existing aesthetic tradition so that gradually geological accuracy became an important criterion for the artist, it is worth stressing that the nineteenth-century geologist was treading the same or broadly similar paths to those that the traveller in search of the sublime and picturesque had trodden in the previous century. The twenty-one year old Charles Lyell, on vacation in Switzerland in 1818, visited Schaffhausen, the Mer de Glace and all the other classic sights of the grand tour at which Horace Walpole and Thomas Gray had marvelled in 1739 and Turner had sketched in 1802. (15) At Schaffhausen, Lyell visited the castle of Lauffen and looked at 'a beautiful landscape of the Fall', with the aid of a camera obscura (a device frequently employed by eighteenth-century painters of views). It is only after he has made observations on the scene that he mentions 'the rock of limestone, and I found some fossil shells in it'. (16)

The ability of artists like John Martin or Turner to suggest cosmic infinitude and the aesthetic grandeur of the earth's surface must have profoundly affected geologists. Notice, for example, not only Turner's consummate skill in the Liber studiorum engravings, but also the way in which he uses natural features to reinforce narrative content and spiritual values in Ulysses deriding Polyphemus - Homer's Odyssey (1829) where two violently lit, gigantic hollowed out arches, derived from studies made in the Bay of Naples the previous year, are placed midway between the two vessels. (17) John Martin's pictures use quasi-geological structures for supernatural effects, and it was doubtless Martin's vision of a preternatural landscape infused with protean energy that Gideon Mantell had in mind when he commissioned Martin to design the frontispiece to The wonders of geology (1838).

The pictorial language used by Hugh Miller reflects precisely that sense of awe before the indefinable that permeates Martin's vast canvases. Attempting to offer explanations about geology, Miller is himself inspired by traditional landscape imagery and the aesthetics of the sublime: 'the sublime prospect presented to the geologist as he turns him towards the shoreless ocean of the upper eternity . . . promontory beyond promontory. . . island beyond island - stretch out in sublime succession into that boundless ocean of eternity, whose sunless irreducible area their vast extent fails to lessen by a single hairbreath . . . ' (18)

When the British Association for the Advancement of Science visited the limestone caverns of Dudley in 1849, the Illustrated London News devoted a page to describing the site and recording the lecture delivered through a speaking trumpet by Sir Roderick Murchison, who, dressed in a 'picturesque and striking' costume of shepherd's plaid and Tyrolean hat spoke on the Dudley formation. Afterwards, we are told of a veritable Martinian spectacle in which:

> red and blue fires were lighted at various parts of the caverns, the effect of which was striking and magnificent in the extreme, and drew forth shouts of admiration from the crowds who thronged the caves; and, as each successive blaze revealed the extent and form of the place, lighting up the projections and angles of the rocks, scenes of indescribable grandeur were produced. The visitors who had arrived earliest at the caverns then retired to

make room for others who could not till then enter; and all strove to get into the fresh air from the sulphureous vapours arising from the burning of the coloured fires. (19)

Such sharing of language and imagery is to be expected since the eighteenth century recognised no dichotomy between art and science; both were directed towards enlightenment. This interdependence lay behind the manifest aim of The Artist, a journal edited by Prince Hoare between 1807 and 1810 which provides an apt illustration of the aspirations of early nineteenth-century men of culture and learning:

> It is the design of The Artist to seek professional information on the subject of the liberal arts in the most distinguished sources of his country, and to present their recondite stores in familiar garb to his readers. With these offerings he proposes to connect accounts of the modern improvements in science, and such observations on them as experience and equally appropriate study can best supply. With this view, he regards every adept of Art or Science under the general description of an Artist, or the active student of Nature and Science; for the practice which renders science useful to life, what is it but Art ? (20)

The easy absorption of speculations, discoveries and researches of geologists in the first half of the century into existing aesthetic traditions made it more difficult for the Victorian art world to accept the authoritative views of Sir Charles Lyell even after they were widely known. The popular Art Journal, first published as the Art Union in 1839, carries not a single reference to geology and little discussion of other sciences until, in an editorial of 1854, it announced patronisingly that: 'Both abroad and at home the men of science would appear to have been reposing. The astronomers have been adding to the number of smaller planets, but these discoveries are so numerous that they cease to interest us.' (21)

The following year, 1855, the Journal carried a short anonymous article entitled 'Geology; its relation to the Picturesque' in which readers leaving town to spend summer in the country are advised that they might find pleasure in the study of geology: 'This science is by many supposed to teach little more than a knowledge of the varieties of rock formations, to deal with a few dry details connected with earths and minerals, and, perchance, to develop a few curious matters, generally regarded however as rather speculative, in relation to remote ages of the earth's history.' After this apologetic beginning, the writer goes on to the premise that 'all beauties of landscape' are 'completely dependent' on 'great geological phenomena'. Working on the assumption that the sole object of the landscape painter is the literal transcription of the seen and understood world, the writer briefly discusses the differences between landscape in different geological areas. (22)

In 1862, a hostile review of *The student's manual of geology* appeared in the *Art Journal*. 'A love of what is above the surface of the earth rather than of what is beneath it, must be our confession', the reviewer begins, and concludes with the Kingsleyian assertion, 'next to astronomy, there is no science which has revealed so much of the power, wisdom, and goodness of the creator of the world [as geology] .' (23)

Then in 1863, the *Journal* carried five long articles by Professor Ansted (24) on science and art subdivided into sections on water; plains; tablelands, hills and valleys; clouds, air and atmospheric meteors; mountains; and, finally, 'on the General relation of Physical Geology to the Progress of Landscape Art in various countries'. All the articles contain a great quantity of factual and scientific information, but the final essay is in many ways the most interesting as it puts forward positive notions about the relationship between landscape painting and geology on a more than merely 'how to do it' basis:

> Until some interest was felt by the general public, and by educated people, in pursuits out of which have arisen the science of Geology, and various departments of physical geography, there was really no taste for landscape Art among civilised nations. (25)

The author does not suggest even an approximate date for the rise of landscape art but it seems fair to assume that, since all his examples from painting are taken from a period after the first thirty years of the nineteenth century, that he is thinking of a post-Lyellian age, thus ignoring Rubens, Claude Lorrain, Salvator Rosa, Richard Wilson, and Thomas Gainsborough, not to mention a host of other earlier landscape painters. Turner's skies and coastal views, Stanfield's cliffs and water, the trees of Birket Foster, the cornfields of Linnell, the rocks of Edward Lear; these present the standard of excellence because, it is suggested, they studied the laws of nature. In other words, Ruskinian truth to nature is the one and only criterion by which landscape painting can be judged. (26)

Considering the examples cited, one is drawn to conclude that there is some confusion here between a laudable desire that the scientific bases of landscape be understood and an admiration for technically highly-finished landscape painting. The omission of Ford Madox Brown and the Pre-Raphaelite *avant-garde* is not, perhaps, surprising. Despite the innovatory nature of their representations of the external world, the Pre-Raphaelite Brotherhood was always more genuinely interested in art and historicism than nature. Most of the artists mentioned are precisely those whom Ruskin praises in *Modern painters* and *Academy notes* for their attention to detailed execution and truth to nature. Constable, who refused to acquiesce to the demands of the Establishment for finicky finish, is not mentioned.

How the geological-artistical school of criticism reacted to Constable can be judged from a remark made by Ruskin's protégé, John Brett: 'Constable took so superficial an interest in nature that he never took any pains to study her laws. All he cared for was to get rough and hasty resemblances of places that he liked to frequent in early life, especially when the weather was showery and squally.' (27) This, of course, is far from the truth. We now recognise the efforts that Constable made to master the causes of natural phenomena in his landscape art. His meteorological studies of the 1820s are well known but it is worth drawing attention here to the interest that Constable showed in geology, especially in 1833 when he made notes on a well-sinking operation on Hampstead Heath and passed them on to his son, John Charles, who was engaged in scientific studies. (28)

To return to Professor Ansted's article, we must notice that the author's historical hypothesis is expanded to embrace a geological theory of style. The German eighteenth-century writer, Winckelmann, whose works, translated and distributed widely, became extremely influential throughout Europe, had expressed the belief that the climate of Greece had produced perfect manly beauty which in turn resulted in the serene grandeur of classical art. (29) Given the English climate, we in this country could never hope to establish a worthy school of History painting. Since the foundation of the Royal Academy in 1768, landscape had been regarded as an inferior branch of art in comparison with History painting, according to the scale of values established by Reynolds in his Discourses. Professor Ansted adapts Winckelmann's climatic theory of art, suggesting that the artist who lives among mountains 'will acquire an eye for bold outline', while he who lives on the plain 'will chiefly appreciate and represent, the details of composition'. Having made this assertion, all that remains for the author to do is to establish that the British Isles offers a great geological variety of landscape and we have a positive theoretical basis for believing in the vitality of native British nineteenth-century landscape painting. This the author proceeds to do. Now, at last, the science of geology had provided the means for an absolute frame of reference for the landscape painter and, through the association of this lowly art form with a spectacular new science, landscape might become as elevated and didactic an art form as History painting. Since the British excelled at landscape painting, our national cultural pride might thus be saved:

> Sacred and historic pictures, and even paintings less serious but equally human, are now fully recognised as teaching important truths; and volumes have been written to explain them . . . But landscapes also teach lessons, and may be suggestive of much that

> is valuable. It is to render these lessons more true and more impressive, and to idealise representations of nature either inanimate or clothed only with vegetable life, that Science steps in and assists Art. A landscape that is the result of some study of nature's methods and laws - that is a true representation of what the intelligent and instructed artist can teach - is, in this sense, a lesson; it affords insight into nature's ways, it points out and directs attention to conditions of the structure of rocks . . . (30)

Although he mentions the union 'that cannot be dissevered existing between the external world and that inner world . . . that makes up the individuality of every one amongst us', (31) Professor Ansted ignores the tricky questions concerning the symbolic value of landscape or the effect of introducing human activity into accurate pictorial transcriptions of landscape. Both these questions troubled Ruskin and his followers.

Reviewing John Brett's Val d'Aosta (R.A. 1859), Ruskin drew attention to a theory of historical landscape which Ansted undoubtedly had in mind when writing for the Art Journal four years later. This was not concerned with Homeric epic scenes in a landscape setting (32) but with pictorial representations of the environment in which the physical history of the landscape was manifest. 'Yes, here we have it at last - some close-coming to it at least - historical landscape, properly so called - landscape painting with a meaning and a use . . . Historical landscape it is unquestionably; meteorological also . . . ' (33) Yet, having extravagantly praised the picture, Ruskin then proceeds to find fault with the artist's neglect of sentiment and emotion. It is insufficiently poetical. It is, as a writer later described Brett's A morning amongst the granite boulders (1873), mirror's work not man's. (34)

The Art Journal enjoyed a wide circulation and Ruskin was the most influential single writer on art of his age. Moreover, the Victorian artist was expected to be a cultivated and knowledgeable member of society, a responsible educator and guardian of public taste. There can be no doubt, in view of this, that by the middle of the century artists were generally in possession of at least the sort of basic knowledge about geology available from these sources. Two specific examples help to illustrate this. Thomas Seddon, writing home from Marseilles on his way to the East in 1853, was able to convey the predominant colour and composition of the landscape he wished to describe by reference to its geological structures as well as to its appearance:

> Avignon. ... The hills around are formed of a white calcareous rock, very broken, and almost bare - their grey hue only broken

by a little dusty mint, and a few prickly herbs, with some olives and pointed cypresses, and stunted pines, wherever a ledge or hollow allows a little soil to lie. (35)

Edward Lear, sketching in the Middle East in 1854, records not merely the colour of rocks but also their geological composition as annotations to his drawings. (36)

However, considering the writings on geology that we have examined from the *Art Journal* for the period 1848-63 and some of Ruskin's comments, it becomes apparent that geology did not always open up new dimensions of knowledge and experience to the landscape artist. It often reinforced academic attitudes to art because it lent itself to geographical and historical theories about the progress of art and provided an absolute standard analogous to that of the classical nude for the History painter. Study of the nude depended on knowledge of anatomy. It is interesting, therefore, to note that writers on geology used anatomical metaphors when describing their own scientific activities. (37)

Discussion on landscape and the environment in the 1750s and 1760s was controversial but it embraced a considerable variety of viewpoints ranging from the sharadwadgi of Sir William Temple to the purely topographical country-house view, and from Italianate vistas to Gainsborough's *Cornard Wood*. (38) One effect of geological science on landscape painting in the 1850s and 1860s was undoubtedly very restrictive. The popular appeal of Lyell and Ruskin must, to some extent, be responsible for the suppression of elements of fantasy in landscape painting. The acceptable norm in landscape painting at this period was a view in which are transcribed all natural features. The alternatives were not regarded as real pictures:

> Drawing and colour, without reference to air, may succeed in producing an effect adapted for a Chinese exhibition, and might be adapted to the oriental taste of the Greek church, but they would certainly not be admitted as producing a picture in the ordinary sense of the word . . . (39)

Geology opened up new areas to the landscape painter that had hitherto been ignored. To Gilpin, the South Downs were boring (40) but to an artist and public possessed of some knowledge of their history and freed from restraints imposed by eighteenth-century expectations of picturesque beauty, they became interesting. Yet, despite this widening of the horizon, landscapes were still graded; now according to a system dependent on geological instead of aesthetic classifications. Truth to nature did not mean a realism which sought to defy the inevitable process of choosing. Ruskin was surprised that Brett had made such a good job of a chalk flint landscape in *The stonebreaker* and hurried the artist off to the granite alps as quickly as he could. (41)

Another problem which attention to geology posed for the artist was that, deprived of the traditional schemata of landscape painting - repoussoir tree, recessional perspective carefully marked out, dark foreground, light middle ground, precisely located figures, and so on, - the artist found it difficult to indicate scale. There was a feeling that the vaster the area depicted, the vaster the canvas had to be. Thus Ruskin found himself complaining that 'there is no occasion that a geological study should also be a geological map' and thought that Brett's Val d'Aosta, which he possessed, 'would have been more precious . . . if it had been only half the Val d'Aosta'. (42)

The prime concern of landscape painters in Western Europe had been the illusion of space. Now that the substance of the earth was more important than the representation of distance, it was difficult for the artist to indicate a total scale. It is impossible to tell whether some of Brett's, Lear's or Ruskin's pictures depict hundreds of feet or one foot of rock. In the hands of a highly imaginative artist this circumstance could, itself, be exploited. This is precisely what William Dyce does in his painting of David with his harp in a geologically convincing Scottish landscape. (43) The doubt about scale is not acute here because the human figure provides some standard by which to measure but as there is only one figure and no real indication of distance we do not know his height and therefore are not able fully to 'read' the scale of the landscape. The degree of doubt that remains reinforces the sense of vulnerability and isolation of this figure.

By 1850 geology seems to have led us to an impasse in which the painter, vying with the photographer, produces increasingly proficient but sterile images of his environment. Yet a few artists were able to move beyond appearances and the daunting authority of scientific laws and theories about progress. They saw that geology could provide the artist or indeed the writer not merely with the means to the sort of stunning pictorial clarity which Ruskin admired in Brett's Val d'Aosta, but also with visual metaphors for metaphysical concerns centred on man's knowledge of himself and the world about him. Writing on the subject of bribery in Past and present (1843), Carlyle was able to exploit a general interest in geology in a highly effective way to expose what he regarded as the moral degeneracy of his age:

> In sad truth, once more, how is our whole existence in these present days, built on Cant, Speciosity, Falsehood, Dilettantism; with this one serious veracity in it: Mammonism! Dig down where you will, through the Parliament-floor or elsewhere, how infallibly do you, at spade's depth below the service [sic], come upon this universal Liars-rock substratum! (44)

The artist was able in a similar way to use geology metaphorically.

One painting by John Brett and two by William Dyce illustrate the effectiveness of this imagery.

John Brett (1831-1902) was described by Ruskin as 'one of my keenest minded friends'. (45) He was a remarkably versatile man who published papers on astronomy, participated in an expedition to Sicily to observe a solar eclipse, and became a Fellow of the Royal Society. (46) He built himself a house in 1877 which, in design, anticipated Bauhaus functionalism, and he became a highly skilled photographer. The latter interest, and the increasing prosperity which enabled him to take his large family on long voyages around the coasts of Britain in his yacht, had a detrimental effect on his art and the repetitive coastal scenes which he exhibited year after year towards the end of his life (probably painted from photographs) lack originality and vision. However, before quarrelling with his mentor, Ruskin, over a scientific matter in 1864, (47) Brett produced one memorable picture, The stonebreaker, exhbited in the Royal Academy in 1858. (48)

The picture comprises a view of Box Hill, Surrey, with a young boy in the foreground who is breaking up stones whilst his puppy playfully chews a cap nearby. It is summer and the shadows indicate that the time of day is late afternoon.

A stonebreaker in chalk country is very likely to find fossils and, therefore, the nature of the activity portrayed by Brett establishes a speculative, theoretical dimension to the picture. It is, after all, no coincidence that the pioneer geologist, William Smith (1769-39), was apprenticed to a surveyor, and that Hugh Miller (1802-56) first became interested in geology when he worked for a stonemason and observed ripple marks on the bed of a quarry. (49) Stonebreakers must have been amongst those who supplied specimens to fossil shops like the one at Lyme Regis. By portraying a stonebreaker, Brett not only presents a figure that fulfils the demands of picturesque landscape but implies, on the one hand, the whole topical question of geology and, on the other, the pressure of social issues like poverty and labour.

The boy's occupation immediately arrests our attention. Two classes of people broke up stones with hammers in early Victorian England, stonebreakers and geologists. Stonebreakers were among the most wretched and poverty-stricken people in any rural community since stonebreaking was the task given to men and boys who were on parish poor relief. Not only were they compelled to suffer the stigma of the workhouse but their degradation was on public view as they sat working by the highway. Geologists were, generally, from a wealthy and highly educated class. The Reverend Adam Sedgwick recounted how a lady who saw him searching for fossils among some stones laid by the road mistook him for a stonebreaker. She gave him a shilling in gratitude for information about the locality and out of pity for his poor state.

> Next evening, to my great amusement she came to dine at the house where I was staying. I recognised her at once, but she did not know me, in my altered dress. She was visiting Wales for the first time, and was full of enthusiasm for the scenery and the people. 'They are so obliging, and so communicative', she said; 'only yesterday I had a long conversation with an old man who was breaking stones on the road. He told me all I wanted to know, and was so civil that I gave him a shilling.' I could not resist the pleasure of saying, 'Yes, Ma'am, you did, and here it is!' (50)

Stonebreaking was an occupation for the very young or for those in extreme old age. The existence of this youthful stonebreaker implies, therefore, a whole lifespan. His youthfulness raises for us a mirror image of old age. We recall Landseer's picture of 1831 in which a young girl brings lunch to her stonebreaker grandfather (51) and Courbet's _Stonebreakers_ (1851) (52) in which an elderly man and a very young boy work side by side. Henry Wallis's _Stonebreaker_ (1857) (53) is a Carlylian figure who has actually died with his hammer in his hand. The rosy-cheeked health of our stonebreaker simply serves to reinforce the inevitable sense of man's life-span which is so very brief in comparison to the only recently understood geological time-span, evidence of which is seen in the 'stony forms of the dead existing millions', (54) lying behind and before his solitary figure.

If we compare Brett's landscape with an eighteenth-century painting attributed to George Lambert of the same view, we may observe a number of differences. (55) By 1858 Box Hill had long been a famous landmark, partly for the extensive views it offers and partly on account of the antiquity of the plantation of box trees which used to cover its sides. Towards the end of the eighteenth century the trees were sold and the cutting of much of the timber accounts for some of the differences. (56) Lambert's landscape, however, shows greater and different evidence of habitation. In it we can see the amateur sketching, the huntsman resting, the landowner inspecting his labourers who are cutting the corn. The nature of the human activity establishes this landscape as an ordered and cultivated scene reflecting the ordered and cultivated society whose representatives are active within it. In Brett's picture we have, instead, a large-scale single figure in the foreground and a huge heap of flints, depicted in minute detail. The juxtapostion of youth and geology concentrate the attention on human and universal time-spans.

If we examine the background of Brett's picture we find we can just discern midway between the boy's head and the tree, the London to Shoreham railway which runs through the gap cut in the chalk downs by the River Mole. The Victorian ideal of progress which the railway embodies is endorsed by the existence of the stonebreaker whose labour helps to make roads.

Stonebreakers were among the poorest labourers in Victorian England. Representations of them in literature and art habitually present miserable downtrodden specimens of humanity. (57) Brett's youthful and healthy young man is the antithesis of this and the total mood of the picture, reinforced by sunshine, pet dog, bright neckerchief, the bird which sings on the tree-top, and the embodiment of progress in transport by road and rail establishes a memorable note of optimism. Reading of Tennyson's fearfulness in In memoriam and recalling Ruskin's well known cry, 'if only the Geologists would leave me alone, I could do very well, but those dreadful hammers! I hear the clink of them at the end of every cadence of the Bible verses', (58) it is easy to conclude that acknowledging the reality of geological time was a depressing business for Victorians. In fact, implied in geological time is the cyclical process of renewal. The dead tree trunk behind the young stonebreaker bears a new branch, new growth. London is twenty-three miles away, several hours of travelling time. Far from any reminder of city filth and degradation this rosy-cheeked young man breaks his stones in tranquil, unspoilt nature, in a land which bears all the signs of present repose, former disturbance and future improvement for the human race.

The effectiveness of The stonebreaker depends not only on its iconography but also on its mode of execution. Brett, a follower of Pre-Raphaelite techniques (59) involving minute detail and brilliant colour, was praised by their champion, Ruskin, for going beyond anything the Pre-Raphaelites had yet done in some points of precision in this picture. (60) The treatment of detail speaks for itself, but Brett's analytical examination of the landscape is worthy of comment. Short of an aerial photograph, I imagine it would be difficult to find a representation of a given landscape that more clearly revealed the underlying structure and its geological history. Brett left no verbal account of his intentions in The stonebreaker but two passages from his unpublished writings seem particularly relevant to a study of this painting. Brett was keenly aware of the transitory nature of human experience and, writing from Farnhurst, Sussex on 13 January 1853, he meditated how great men 'make a little sensation and then drop out of sight'. He thought it would be better to be remembered in the hearts of a few, 'or even to be known appreciated loved only by one pure spirit - there would be some hope of success here - the heart's blood would not flow away into an ocean or over a heap of flint stone'. (61) Later, in a note in an album dated 30 January 1862, Brett described the appearance of human flesh in a passage which reveals the considerable powers of observation and analysis that the artist possessed:

In flesh the red element is common to lights and shades. The semi opaque yellowish cuticle is of equal transparency and polish everywhere, but is unequally corrugated according to its tightness or looseness over angles, being fully stretched, its surface reflects most perfectly, over soft muscles and in hollows it lies like a ploughed field still equally polished but the surfaces being broken up . . . but at deep angles they produce almost the depressive effect of ground glass. In the recesses of the corrugation the maximum transparency of the article is appreciable and the blood colour shows itself; on the ridges well surrounding lights are reflected producing either a confused greyness or whiteness according to the degree of colour of the light. (62)

William Dyce, who died in 1864, was a prominent figure in the Victorian art world. A Royal Academician, ecclesiologist, and expert on art education, he was very much more of an intellectual than John Brett. He painted a limited number of pictures but nearly all of them are characterised by complex imagery drawn from the natural world. _Pegwell Bay: A recollection of October 5th 1858_, (63) which Dyce exhibited in 1860, is arguably his most popular and best known work. It has always provoked a response which, though not consistently favourable, has been remarkable for its intensity. In 1902, D. S. MacColl thought of the artist 'dwelling on the shores of chalk that he has painted in a desperate pallid gleam of imagination' and felt that the painting was pervaded by gloom: 'It is as if man had come to the ugly end of the world, and felt bound to tell.' (64) A more recent critic described _Pegwell Bay_ as 'at once simple and wise . . . perfect in execution and yet filled with an unequalled personal feeling', and wrote of 'the crystal-clear sky where the late afternoon light gives a curious feeling of waiting, of unease mingled with joy, of a calm solemnity which was for the artist the dominating emotion of this fifth of October which he has so magnificently perpetuated'. (65)

What are the ingredients in this painting which have provoked such an emotional response? Dyce portrays the shore of Pegwell Bay near Ramsgate with the tide out on a cool Autumn day. A few people wander about the shore, fishing in rock pools, leading donkeys, catching the shrimps for which the bay was famous at the time, and, in the case of the male figure at the far right of the canvas, carrying artist's materials. Perhaps this is the artist himself since this is a very autobiographical picture and we know the foreground figures to be (from right to left) the painter's wife and her two sisters and one of the artist's sons. (66) In the sky Donati's comet, first observed on 2 June 1858, can be clearly seen (though not, unfortunately, in a photograph).

Whilst we have no sketches and little documentary evidence to bring to bear on an examination of Brett's _The stonebreaker_, the reverse is the case with _Pegwell Bay_. Whether or not Dyce accurately depicted the geophysical features of the site and whether or not he used photographs to assist him are matters that have been hotly disputed. James Dafforne, writing in the artist's lifetime, denied that the painting was based on a photograph. (67) The artist's cousin, the eminent Shakespearean scholar Alexander Dyce, suggested that William Dyce had embellished the features of Pegwell Bay and misrepresented the colour of its cliffs. (68) In fact, Dyce had been a close friend of David Octavius Hill since the 1830s and certainly knew about photography. It is clear, however, from comparing the watercolour sketch, dated October 1857, (69) on which Dyce based his picture, with the finished work that he was not primarily concerned with a literal transcription of his surroundings. Details like the condition of the fence at the right and the presence of figures vary but there are also major differences in the geological structure of the cliff. In the finished painting the caves have become shallower, the cliffs higher, and a prominent gulley has replaced the central cave. Most significant is a change in mood. The emotional effect of the sketch is concentrated in the power of the landscape. The figures in the painting fail to weaken the dominance of the natural surroundings but they introduce an element of dream, of nostalgia, and of the consciousness of time passing.

The full title, _Pegwell Bay: A recollection of October 5th, 1858_, stresses the quality of memory, and the date of the sketch (1857) suggests that the memory which is celebrated in the picture, exhibited in 1860 and presumably painted in 1859 or 1860, is of a time more distant than is indicated in the title of the painting. In his insistence on recollection, Dyce was at variance with the practice of Pre-Raphaelite artists who spent hours ensuring complete visual accuracy. Dyce succeeds in rendering a landscape with extraordinary clarity and an air of exactitude but his main concern is with the intellectual response to this particular environment. _Pegwell Bay_ is a painting about time, explored through an image of a particular moment in time.

As a young man in Aberdeen, the son of a Professor of Medicine at Marischal College, Dyce had studied natural philosophy and wrote, in 1832, a prizewinning essay on electro-magnetism. Moving to Edinburgh in the early 1830s, he could not have failed to be aware of the Werner-Hutton controversy that had been raging there. Dyce's friends were academics and men of science. His nephew, James Clerk Maxwell, was to become one of the greatest scientists of the century. After moving to London, Dyce became Professor of Fine Arts at King's College in 1844 and, although his activities there seem to have been minimal, through his appointment and through his friendship with men at the newly established Chemical College, he is likely to have been cognisant with at least the general tenor of scientific thinking at this time.

The original sketch for Pegwell Bay was painted, as we have seen, the year before the date which appears in the title of the exhibited work. Dyce's inclusion of a date in the title does not spring from the desire to record accurately and scientifically weather conditions on a particular day which motivated Constable, William Mulready and other nineteenth-century English landscape painters. (70) There is scarcely a perceptible movement of sea or cloud in Pegwell Bay and no breath of wind stirs the ladies' shawls. There can be only one reason for the date given by the artist, and that is Donati's comet. Dyce chose to paint the day on which the comet appeared at its most brilliant and when its development was being recorded by astronomers all over Europe. (71) On 2 October 1858, The Illustrated London News printed an account of the progress of the comet from its own correspondent:

> The comet becomes more and more brilliant on each evening that it can be seen and bids fair to exceed all those which have been visible since the wonderful one of 1811, of which we have heard so much . . . As a short sketch of the history of the present remarkable comet, we may state that it was discovered by M. Donati, the astronomer, at Florence, on June 2 of the present year, when it was of extraordinary faintness . . . It may be stated for certain . . . that the comet is now passing directly to the star Arcturus, and, as the comet and this star are the brightest objects in the western heavens during the evenings, its course on October 5th may easily be traced. At six pm of October 5 it will be a very little to the right, and a very little to the south of Arcturus, and we hope our readers may have a clear sky to witness the conjunction of two such bright objects, which will be separated by but a very short distance. (72)

Pegwell Bay is, therefore, an extremely topical painting. It could be that Donati's comet in the painting simply pinpoints the day and the time more accurately. However, astronomy is not the only science which plays a part in the imagery of the painting. Tennyson described astronomy and geology as 'Terrible Muses' (73) and Hugh Miller viewed astronomy ('the depths of heaven') and geology ('the bowels of the earth') as natural polarities. (74) It is, surely, no coincidence that astronomy, represented by the comet, is accompanied in Dyce's painting by her sister Muse, geology, apparent in the crystalline clarity of the chalk cliff and the shell-strewn beach. The artist's relatives are seen in microscopic detail, strung out like pebbles in a stony land where all activity is desultory and meaningless, whether that of the donkey-keeper, the artist, or the boy looking for shells or fossils. Humanity is threatened on either side by the 'Terrible Muses' of which it seems unaware. The figures do not eye astronomy or geology or even each other. The painter transfixes time: a particular moment on a particular day in 1858. The comet, however, dominates that late afternoon and represents a system immensely distant in space ('on Oct 2nd . . . it is sixty-two million miles from us') and incalculable to any degree of accuracy in time ('it would seem that this comet will return in 2100 years time; but with comets of this period . . . calculators cannot be certain to a few centuries') (75). The 'Terrible Muses' deny the validity of a single human life and, even more, of a single day. Nevertheless, sea-shell gathering is - it would appear - the only reality for these people in this place on this day.

Significantly, Lyell mentioned astronomy in order to indicate that _his_ subject, despite its immensity, is microscopic in comparison with that of the astronomer:

> the geologist may admire the ample limits of his domain, and admit, at the same time, that not only the exterior of the planet but the entire earth, is but an atom in the midst of the countless worlds surveyed by the astronomer. (76)

A man of Dyce's intellectual standing would certainly have been familiar with Lyell's theories. Tennyson was early in his reading of Lyell, whose views prompted some painful speculations in _In memoriam_. (77) Lyell's _Principles of geology_ was widely known and accepted by the 'fifties. A letter from Dyce to his brother-in-law, written in 1860, the year in which _Pegwell Bay_ was exhibited, testifies to the artist's knowledge of geology and his understanding of erosion or 'the process of disintegration':

> The only place I have seen in Scotland which reminds me of the very wild parts of North Wales is Glen Rosa in Arran . . . but there are a hundred such places in Wales . . . and the mountains are generally more rugged, stony and precipitous, more awful and terrible looking than anything I know of in Scotland . . . One great cause of the beauty of the Welsh mountains is, I think, to be found in their geological formation. In Scotland the granite mountains, by the process of disintegration become rounded and their asperities smoothed down . . . but in Wales, the material being slate rock, it does not crumble like granite into dust or sand but splits and tumbles down in huge flakes which leaves the peaks from which they have fallen as sharp and angular as if they had never been acted upon by the atmosphere at all. (78)

During his first and only visit to Wales in 1860, Dyce painted a picture in which he reiterates the imagery of time which is central to _Pegwell Bay_. However, _Welsh landscape with two women knitting_ (79) makes no concession to nostalgia, to memory, or to affection - that is, to those elements which man uses as a weapon against time and which Dyce's contemporary, Tennyson, exploited so effectively in his poem _In memoriam_. _Welsh landscape_, which is harsh in colouring (probably due to the influence of Pre-Raphaelite art) is entirely lacking in any sort of pastoral lyricism or subjective evocation of mood. It is as severe a statement as one could expect to find in a painting which draws on the traditions of naturalistic landscape and genre. It cannot, however, be explained simply as a realistic representation of Welsh natives in their natural habitat. (80) The two women, surrounded by ancient rock formations, are knitting. Their activity, like Penelope's weaving or the ominous knitting of the black-clad old woman at the beginning of Conrad's Heart of darkness, suggests the inexorable progress of time. The woman on the right is young and beautiful. The face of the woman on the left is wrinkled with extreme old age and yet her age is as nothing compared to the age of the rocks on which she is seated. Above the two Welsh women a sickle moon suggests the cyclical routine of the universe beyond man's control. In Welsh landscape with two women knitting the only sign of habitation is a small hut in the background. This painting and Pegwell Bay differ from the traditional English inhabited landscape. Pegwell Bay lacks the fishing boats and fishermen which typify the genre and which Dyce himself introduced into earlier seashore scenes. (81) Nor is it characterised by the frenetic holiday fun which we associate with W. P. Frith's Ramsgate Sands (82) and other paintings celebrating the Victorian seaside. Behind Frith's holidaymakers is a substantial and dignified terrace of houses. Man has put this beach in its place. Pegwell Bay is almost as devoid of human habitation as Welsh landscape. Only the roof of a house is seen above the cliffs which are threatened with erosion and loom ominously over a pebbly beach, itself witness to the way in which geological time grinds rocks into stones. Unlike the authors of guide books who extol the beauties of Pegwell Bay viewed from its commanding cliffs where, in the 'fifties a stately row of marine villas was erected, (83) Dyce takes up his position at the foot of the chalk cliffs. The human figures consequently appear completely vulnerable. They are neither dishevelled by wind nor dirtied by sand, but their clothes are hardly sufficient shield against the eye of the comet and of the rocks, neither of which they see. Dyce's coherent and sensitive appraisal of human perplexity before the beauty and the threat of time made discernible in nature is equalled in its period only, perhaps, by Matthew Arnold's Dover Beach with its 'grating roar of pebbles' and its 'eternal note of sadness' at the loss of religious faith and disillusionment with human progress. Arnold's metaphor of ebb and flow is a powerful one. The pictorial artist in the nineteenth century did not work with literary metaphor but through the articulation of the visible world. Dyce in Pegwell Bay and Brett in The stonebreaker eschewed the verbal explanation and long discursive title which are the hallmark of much English painting of this period from Turner to Ford Madox Brown. In these most provocative and questioning of paintings, the artists achieve through primarily pictorial means (the title of Pegwell Bay is, despite its comparative brevity, important) the portrayal of two concepts of geological time in opposition to human, daily time, the one pervaded by a mood of optimism, the other bearing the hallmark of intelligent pessimism.

The stonebreaker and Pegwell Bay are unusual paintings. Executed in a period when speculation and revelation about our physical environment made it difficult for artists to continue to look at nature as a benign reflection of divine power or regard landscape painting as 'the last branch of uninteresting subjects', (84) these two paintings unite an intellectual response with a desire to record. The eye and mind are activated through association in scenes that do not avoid the role played by the visual artist in observing and recording rather than philosophising and discoursing. The locations are precise, the detail faultless, and yet this sort of precision is not an end in itself. By the end of the century, the function of landscape art as information (a tradition the roots of which lie in seventeenth-century topography) was dead. Modern media have made the topographical draughtsman and landscape painter superfluous. The preoccupation with empiricism and with optical truth which we have observed as a major factor in British landscape painting in the 'fifties was finally worked out in the canvases of Impressionist artists in France from the decade of the 'sixties onwards. For a continuation of the tradition in which carefully selected and geologically accurate landscape is used as imagery in philosophically discursive art, we must turn to Max Ernst who created mysterious (and strangely urban) landscapes by taking rubbings from natural substances like tree bark, or, indeed, to Paul Klee, the mathematician, musician and artist who recorded pictorially his simultaneously physical and psychological discoveries about the world around him.

Notes

1 P. Klee, "On modern art", first published 1924; (tr. and intro. by H. Read), London, 1948, p.49.

2 Sir J. Reynolds, "Discourses on art", ed. R. Wark, New Haven & London, 1975, 'Discourse iv, Royal Academy 1771', pp.69-70.

3 For a useful discussion of modes of landscape painting and attitudes to nature in this period see L. Parris, "Landscape in Britain c. 1750-1850", London, Tate Gallery, 1973.

4 M. Rudwick, 'The emergence of a visual language for geological science 1760-1840', Hist. Sci., 1976, xiv, 149-95 (153). An extremely detailed and interesting discussion of the aesthetics of singularity with relation to geological features came to my notice too late for any proper reference in my text. The article is, however, relevant to the subject of this essay: B. M. Stafford, 'Towards Romantic landscape perception:

illustrated travels and the rise of "singularity" as an aesthetic category', Art Quart., 1977, ns i, 89-124.

5 For example, W. Gilpin, "Three essays: On picturesque beauty; On picturesque travel; and On sketching landscape", London, 1792.

6 J.H. Fuseli, "Lectures on painting", London, 1820, p.185.

7 E. Burke, "A philosophical inquiry into the origins of our ideas of the sublime and the beautiful", London, 1757.

8 Rudwick, op.cit. (4), figs. 16, 17.

9 See Gilpin, op.cit. (5), and A. Cozens, Notes for 'Principles of Landskip', British Museum MS, 1888-1-16-9-2 . . . 5.

10 A. Wilton in C. White (ed.), "English landscape 1630-1850: drawings, prints & books from the Paul Mellon Collection", New Haven, Yale Center for British Art, 1977, p.xviii.

11 Brighton Art Gallery.

12 Tate Gallery, London.

13 J. Ruskin, "Modern Painters" (1844), 2nd edn., in "The works of John Ruskin", ed. E. T. Cook & A. Wedderburn, 39 vols., London, 1903-12, iii, 21-2, 32.

14 Illustrated London News, 30 September 1854, p.308.

15 For Gray's celebrated description of the Grande Chartreuse, see "The correspondence of Thomas Gray", 3 vols., Oxford, 1935, i, no. 71, Thomas Gray to Mrs Gray 13 October 1739. For Turner see, for example, "The Mer de Glace, Chamonix" and "Glacier and source of the Arveiron", both watercolours, British Museum.

16 K. M. Lyell, "The life, letters and journals of Sir Charles Lyell", 2 vols., London, 1881, i, 79.

17 National Gallery, London.

18 H. Miller, "First impressions of England and its people" (1847), 11th edn., Edinburgh, 1870, p.309.

19 Illustrated London News, 22 September 1849, p.201.

20 The Artist (papers 1807-10 collected in 1 vol.), London, 1810, pp.10-11.

21 Art J., 1854, pp.367-8.

22 Art J., 1855, pp.275-6.

23 Review of J. Beete Jukes, "The student's manual of geology", Art J., 1862, p.179.

24 David Thomas Ansted FRS., 1814-80, Professor of Geology at King's College, London, and Assistant Secretary of the Geological Society 1844-7.

25 Art J., 1863, p.235.

26 'I shall look only for truth; bare, clear, downright statements of facts; showing in each particular, as far as I am able, what the truth of nature is, and then seeking for the plain expression of it, and for that alone.' J. Ruskin, 'Modern painters', in "The works", op.cit. (3), iii, 138.

27 J. Brett, 'Landscape at the National Gallery', Fortn. Rev., 1895, ns lvii, 623-39 (638). Ruskin made it his responsibility to supervise Brett's sketching holiday in the Alps in 1858 and

to give him a thorough 'hammering'. Ruskin to his father, 26 August 1858, "The works", op.cit. (13), xiv, pp.xxiii-xxiv.

28 Letters of September and October 1833, "John Constable: further documents and correspondence", ed. L. Parris & C. Shields, London, Tate Gallery, 1975, pp.85-91.

29 J. J. Winckelmann, "The history of ancient art" (1764), in D. Irwin (ed.), "Winckelmann: writings on art", London, 1972, pp.104-44.

30 Art J., 1863, p.235.

31 Ibid.

32 The sketching club founded by Thomas Girtin in 1799 hoped to establish a school of historic landscape, the subjects being original designs for 'poetick passages'. See J. Hamilton, "The Sketching Society 1799-1851", London, Victoria and Albert Museum, 1971.

33 J. Ruskin, "The works", op.cit. (13), xiv, 234-7.

34 Art J., 1873, p.236.

35 J. P. Seddon, "A memoir and letters of the late Thomas Seddon by his brother", London & Edinburgh, 1858, p.26.

36 E.g. "Mahatta", watercolour, Brighton Art Gallery.

37 Art J., 1863, p.149: 'Mountains are the . . . skeletons whose more marked ridges and more prominent bones are manifest through all the coating of muscle and flesh and skin and are recognised at once as the foundation of everything that is manly and vigorous in external nature.' H. Miller, op.cit. (18), p.175: 'When Signor Sarti exhibits his anatomical models, he takes up one cover after another . . . until at last the skeleton is laid bare. Let us, in the same way, strip the vast landscape here of its upper integuments.'

38 T. Gainsborough, "The Cornard Wood", 1748, National Gallery, London. 'Sharadwadgi' was a term used in Chinese-style gardening.

39 Art J., 1863, p.193.

40 W. Gilpin, "Observations on the coasts of Hampshire, Sussex and Kent, relative chiefly to picturesque beauty made in the summer of the year 1774", London, 1804, pp.40-1.

41 'Nearly the last stone I should ever have thought of anyone's sitting down to paint would have been a chalk flint. If he can make so much of that, what will Mr. Brett make of mica slate and gneiss?' J. Ruskin, "The works", op.cit. (13), xiv, 171.

42 J. Ruskin, 'Academy notes', in "The works", op.cit. (13), xiv, 293.

43 W. Dyce, "David", c. 1855, private collection.

44 T. Carlyle, "Past and present" (1843), London, 1888, p.218.

45 J. Ruskin, 'Modern painters', in "The works", op.cit. (13), vii, 360.

46 H. H. Tunner, obituary in Monthly Notices Roy. Astron. Soc., 1902, lxii, 238-41.

47 J. Ruskin to D. G. Rossetti, July 1865, "The works", op.cit. (13), xxvi, 494.

48 Walker Art Gallery, Liverpool.
49 H. Miller, "My schools and schoolmasters or the story of my education" (1854), Edinburgh, 1879, p.155.
50 J. W. Clark & T. McKenny Hughes (eds.), "The life and letters of the Reverend Adam Sedgwick", 2 vols., Cambridge, 1890, ii, 573-4. I am grateful to Roy Porter for drawing my attention to this passage.
51 Victoria and Albert Museum, London.
52 Destroyed 1945, known from photographs.
53 Birmingham City Art Gallery.
54 Miller, op.cit. (18), p.82.
55 Tate Gallery, London. The identification of the landscape subject has recently been questioned, but no definitive evidence has been produced to demonstrate that it is not Box Hill. On purely visual evidence this would seem to be the location. See also Richard Wilson's study for a lost painting of Box Hill. The drawing, in the Mellon collection, the Yale Center for British Art, is a wide panoramic view with a group of figures in the foreground and appears to have been taken from the same spot as Lambert's view.
56 F. Shoberl, "The beauties of England and Wales", ed. J. Britton & E. W. Brayley, 17 vols., London, 1801, etc., xiv, 159-60.
57 An interesting corollary to this appears in a recent children's book: E. Fairfax-Lucy & E. P. Pearce, "The children of the house", London, 1968. The book is based on E. Fairfax-Lucy's recollections of his Edwardian childhood. The worst fate that could befall the son of a gentleman who failed to work hard would be to become a stonebreaker. This is the threat that hangs over the head of the son of the house.
58 J. Ruskin to Henry Acland, 1851, "The works", op.cit. (13), xxxvi, 115.
59 For a detailed discussion of this subject, see A. Staley, "The Pre-Raphaelite landscape", Oxford, 1973.
60 J. Ruskin, 'Academy notes', in "The works", op.cit. (13), xiv, 171.
61 John Brett's diary, private collection MS. I am indebted to David Cordingly for drawing my attention to this passage.
62 An album of sketches, private collection MS.
63 Tate Gallery, London. This discussion of paintings by William Dyce first appeared in M. Pointon, 'The representation of time in painting: a study of William Dyce's _Pegwell Bay: a recollection of October 5th, 1858_', Art Hist., 1978, i, 99-103. Further information concerning Dyce and his contacts with scientists will be found in my forthcoming book, "William Dyce: a critical biography", Oxford, 1978, or early 1979.
64 D. S. McColl, "Nineteenth-century art", London, 1902, p.115.
65 M. Brion, "Art of the Romantic Era", London, 1966, pp.95-8.
66 Dyce papers, Aberdeen Art Gallery MS.
67 J. C. Dafforne, 'William Dyce R.A.', Art J., 1860, pp.293-6.

68 A. Dyce, "The reminiscences of Alexander Dyce", ed. R. J. Schrader, Columbus, Ohio, 1972, pp.212-3.
69 Private collection.
70 See Constable's cloud studies in the Victoria and Albert Museum, London, and drawings by William Mulready in the Whitworth Art Gallery, the Univesity of Manchester.
71 Illustrated London News, 16 October 1858, p.349.
72 Illustrated London News, 2 October 1858, p.309.
73 'Parnassus' (1889), in C. Ricks (ed.), "The poems of Tennyson", London, 1969, p.430.
74 Miller, op.cit. (18), p.307.
75 Illustrated London News, 2 October 1858, p.309.
76 C. Lyell, "Elements of geology" (1838), revised edn., London, 1865, p.2.
77 Published in 1850 but written over a period of twenty years.
78 W. Dyce to R. D. Cay, 20 October 1860, Dyce papers, Aberdeen Art Gallery.
79 Private Collection.
80 See L. Nochlin, "Realism", London, 1971, p.124: a 'harmless representation of colourfully costumed natives in their natural habitat'.
81 See, for example, "Culver Cliff, Isle of Wight", watercolour 1847-8, Mellon collection, Yale Center for British Art.
82 1854, Royal Collection.
83 Shoberl, op.cit. (56), viii, 985; and anonymous author, "The land we live in: a pictorial and literary sketch book of the British Empire", 4 vols., London, n.d. [c. 1850], i, 157.
84 Fuseli, op.cit. (6), p.185.

1 William Hodges, "The inner crater of Mauna Loa, Hawaii", oil, Royal Pavilion Art Gallery and Museums, Brighton

2 John Martin, "The plains of heaven", oil, Tate Gallery, London

3 John Ruskin, "Study of gneiss rock, Glenfinlas", pen and wash, Ashmolean Museum, Oxford

4 John Brett, "The stonebreaker",
oil, Walker Art Gallery, Liverpool

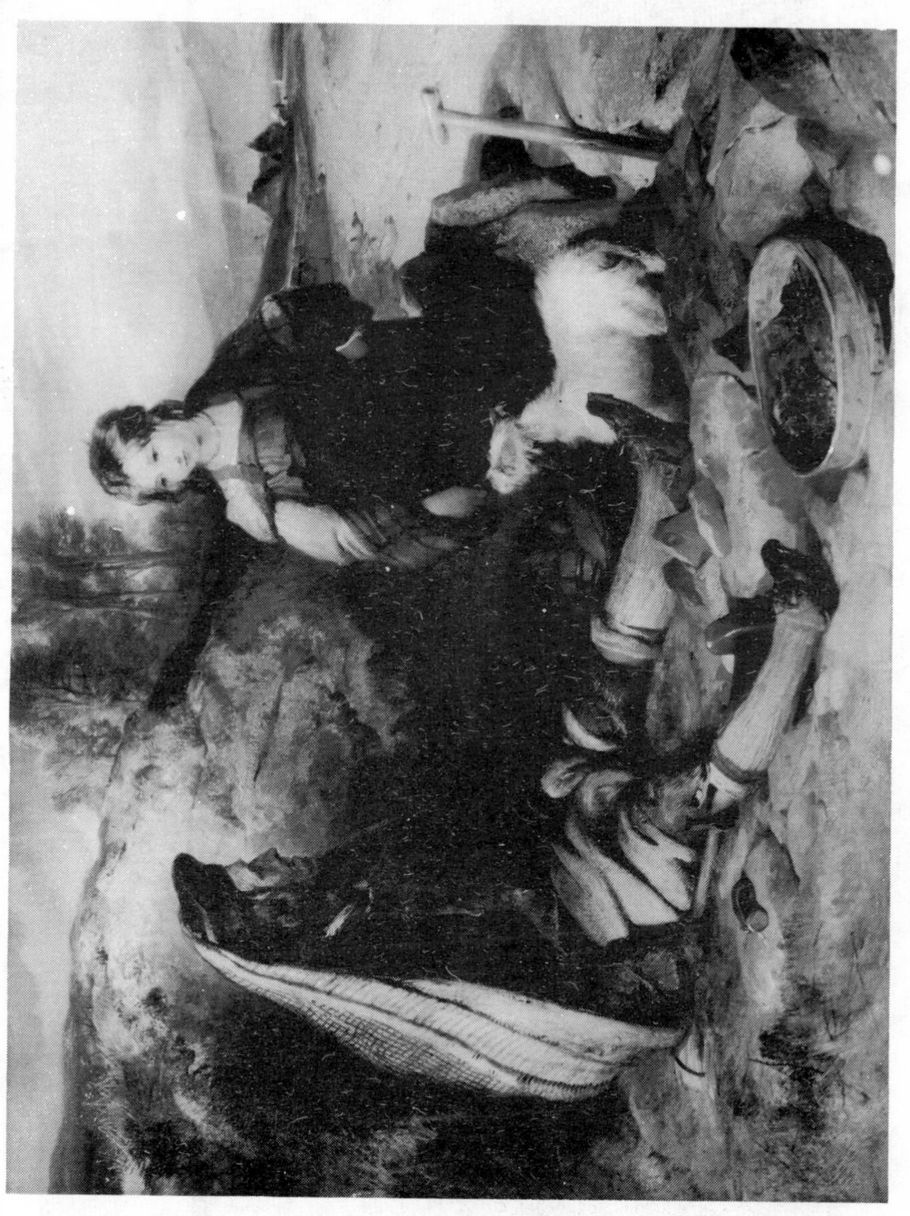

5 Sir E. Landseer, "The stonebreaker",
oil, crown copyright, Victoria and Albert Museum, London

6 George Lambert, "Box Hill", oil, Tate Gallery, London

7 William Dyce, "Pegwell Bay: a recollection of October 5th 1858",
oil, Tate Gallery, London

8 William Dyce, "Welsh landscape with two women knitting", oil, private collection

Earth science and discipline boundaries

Earth science and environmental medicine: the synthesis of the late Enlightenment

L. J. JORDANOVA

> This disease displays a wonderful coincidence in time, with earthquakes, volcanoes, unusual seasons, and the appearance of comets. (1)

Introduction

The connection of events on the earth and in the atmosphere with human ailments was a common theme of the scientific and medical literature of the 1790s. Taking thattheme as a starting point, this essay explores the environmentalism of the late eighteenth and early nineteenth centuries with special reference to the way the earth sciences were used to explain disease.

The environmental sciences - geology, geography, meteorology - were intimately related to the biomedical sciences - physiology, biology, hygiene - as well as to the human and social sciences such as anthropology and epidemiology. Using environmentalism as the common concern of these apparently disparate areas, the paper argues that the synthetic approach of the late Enlightenment had direct social implications. These consisted in the refinement and elaboration of manipulative techniques of managing individuals and groups by an increasingly self-confident medical profession. Furthermore, the synthesis of earth science and environmental medicine, manifested for example in hygiene movements, was present in Germany, France, Britain, and the United States in remarkably similar forms at the same time. The material demands extensive comparative study for both the common features and the differences to be revealed.

My basic concern is to understand the self-consciously scientific attempts to master the laws of the organism-environment relationship which were expressed in the vast body of literature on the subject at the end of the eighteenth century. It may be convenient to distinguish four facets of this literature: the physical, the physiological, the anthropological, and the social, which display increasingly complex assumptions about the interaction between human beings and their environment.

First, the physical environment, as studied by geology for example, offered direct explanations of how external conditions affected living things. This interest in physical conditions, also seen in climatology, meteorology, topography and geology, is well shown in the extensive published descriptions of travels which proposed to explain the forms of life by reference to the environment. There was an explosion of fact collecting, particularly among those who gathered meteorological data, often three times a day, with little theoretical guidance or specific aim. (2) Frequently such people were country physicians and clergymen or members of provincial learned societies. In other cases, the empiricism was part of a carefully organised project for gathering environmental data. Anonymous or untraced correspondents provided large amounts of information to well known writers like Lamarck and Volney. Their work was extremely important in the accumulation of observations which could be used for natural philosophical purposes, but the extent of such scientific activity remains largely unknown.

Second, there was an interest in the physiological details of environmentally caused diseases, seen in the concern with diet, with respiratory problems, and with common diseases believed to be caused by mephitic substances emanating from unhealthy places like marshes and swamps. The medical writing on yellow fever in America during the 1790s was of this kind.

Third, there was an anthropological dimension to environmentalism which took the form of descriptions and analysis of the lifestyle, health and diseases of other cultures, emphasising variation from one part of the world to another. (3) This was clear in the writings of both lay and medical observers of the time, such as Volney's descriptions of the Near East and of America. (4)

Finally, the social approach analysed the impact of environmental conditions which stemmed not from climate but from civilisation itself. Bills of mortality and the conditions in institutions, factories and slums were used to draw attention to the impact of the social and moral environment on human well-being. This approach was developed under the leadership of men like Louis René Villermé (1782-1863), the author of many memoirs on the hygiene of the poor, factory workers and prisoners. (5) Hand in hand with this went an examination of the role of the medical profession in dealing with socio-medical problems, expressed for example in the writings on medical police which suggested the central role of doctors in maintaining public order. (6)

It was in medical writing that these common concerns of the earth and the life sciences were most clearly expressed. The expression 'medical writing' should not be taken to imply that the authors were uniformly medically qualified. In fact, several prominent figures were not, such as Louis Cotte (1740-1815), the meteorologist, Jean-Baptiste Lamarck (1744-1829), the naturalist, Constantin Volney (1757-1820), the linguist and traveller, and Noah Webster (1758-1843), the lexicographer. Whether authors were medically qualified or not, whether they were French or American, certain themes recurred: empiricism, utility, and the management of social problems.

Despite its natural philosophical heritage, the literature on the environmental causes of disease was empirical in orientation. People were exhorted to gather ever more voluminous data; detailed observation - including the study of animal disease, the growth of crops and human health - was the praised and practised method. Direct practical utility was claimed to follow from such work. Social usefulness was thought to derive from discovering regularities or natural laws which led to ways of managing disease and so alleviating misery. It could be as simple as draining a marsh, or more elaborate as in building dwellings on exactly the right spot, or more directly manipulative, as in appealing to people to develop more moderate and regular habits, in eating and clothing for instance, as a means to achieving a healthy equilibrium with the environment. The hope of basing social policies on environmental data was explicit. Writers on the subject revealed their consciousness of themselves as guardians of the people, the well-informed protectors possessing specialised knowledge with which to formulate rules to benefit the community as a whole.

A natural philosophy which stressed the interrelationship between the components of nature, took it for granted that the earth, atmosphere and solar system - in fact all aspects of the environment - affected living things. Increasingly during the eighteenth century it was also assumed that the systematic features of the organism-environment relationship could be understood, and then used to affect it. Learning about nature was thus a step towards controlling it. Human intervention worked within the boundaries set by nature's laws which had to be both understood and obeyed. This emphasis on knowledge and expertise enhanced the importance of a group of people who had both the body of scientific knowledge and the opportunity to apply it to concrete situations: the medical profession. Beneath the environmentalist literature lay a new and expanded role for the physician, suggested by the metaphors of doctors as legislators, judges, and police.

There was another sense in which medicine provided a model of how to conceptualise environmental influence. The study of disease was revealing the ways in which both the physical and the cultural environment affected organisms so that the interaction between life and its milieu was being conceptualised in medical terms. This understanding was applied to the group as well as to the individual. 'Public hygiene' first became an area of study in the late eighteenth century. Social pathology, like individual pathology was formulated in environmental terms. The remedies at both levels were the same, the application of rules of hygiene, administered by the physician, to safeguard the health of society. Thus extrapolations were made from physical to mental and on to social traits, and from individual ills to collective ones. This move was permissible since all these characteristics had their roots in 'nature', and could be explained by the level of 'organisation' the natural object in question displayed. (7) Such a naturalistic perspective had become common enough for it not to seem strange when the physician took both society and the individual as a patient.

This synthesis of earth science and environmental medicine had a sophisticated philosophical ancestry as well as deep roots in practical science and in reform movements. The importance of natural philosophy is hard to overestimate. Sensualist epistemology, Neo-Hippocratism, theories which stressed organic flexibility and dynamism, were aspects of eighteenth-century thought which were closely bound to the scientific study of the environment. Cosmological considerations underlay the ways in which savants of the period understood the impact of nature on mankind and of mankind on nature. (8) On the one hand there was tremendous respect for the potency of natural laws which suggested that obedience to such laws was in the best interests of the human race. Nature was the supreme force capable of unrivalled destruction in earthquakes, tidal waves and volcanic eruptions. On the other hand, the physical parameters which affected the receptive realm of life could be measured, manipulated and understood, making human beings also agents of change. Natural conditions could be altered directly, as in the draining of swamps and the clearing of forests, topics of great interest to hygienists. (9) Furthermore, knowledge and understanding of natural laws allowed people to devise ways to lessen the impact of nature, mostly through correct living.

Two themes at the heart of the life sciences were part of the natural philosophical framework of the late eighteenth century: the notion of adaptation, and the idea that both life and the environment have a history. Although adaptation was part of the 'argument of divine design', it was also central to the more secular, naturalistic approach I am examining. It emphasised the fit between the organism and the environment, the interconnections between the three realms of nature - animal, vegetable, mineral - and the desirability of being well adapted to physical and social surroundings. (10) This was not easy because each individual or nation brought with it a legacy of past habits built up by experience, which had to be moulded to new situations such as emigration. The history of nature was thus important for understanding its present state. The temporal development of life and of the earth could be seen as two parallel although interacting strands. (11) In a similar way, illnesses could be fully understood only by reference to the patient's past, manifested by his or her constitution, while animals and plants had to be studied in the light of information about their changing habitats, such as ancient upheavals on the face of the earth.

It was therefore no accident that history and geography came together in this period. (12) In fact, both history and geography frequently used the language and insights of environmentalism. (13) This happened in several ways; there were not only common ideological concerns which underlay the different environmentalist positions, but there was also a shared body of empirical data which was drawn on for support. Hygienists, especially, touched on some of the most sensitive social and political issues which historians, political theorists and geographers were tackling. Their influence was immense, not only in the information they collected or in the policies they advocated, but also in the images they created about the medical and scientific causes of poverty, (14) and about the way of life of other cultures, past and present. (15)

The proponents of environmental medicine occupied positions of some influence. They were active in reforms of all kinds: medical education, hospital organisation, provision for the criminal, the insane and the poor. They either held government positions or were advisors to the state. They wrote popular and semi-popular works, contributed to encyclopedias and organised journals for the educated layman. (16) The army and the navy had a special interest in fostering such work. The synthesis of earth science and environmental medicine was thus embedded in all aspects of the life of the late Enlightenment from natural philosophy and other intellectual pursuits to the formulation and implementation of social policy.

The concept of hygiene

The concept which, more than any other, suggests the synthesis the present essay seeks to unpack is that of hygiene. Littré defined it as that part of medicine which deals with the rules for the preservation of health. (17) The term hygiene denoted the concern for the well-being of animals and human beings expressed through the search for principles to ensure health which derived from an understanding of the impact of the environment. Individual and collective habits became objects of great interest to medicine and subject to general social scrutiny.

Although there was no shortage of treatises on the subject in the English speaking world, for both American and British authors hygiene was an especially French undertaking. (18) Thus we can emphasise the French literature, but safely extrapolate our findings to other contexts. The synthetic nature of hygiene may be seen in the thirteen volume medical section of the _Encyclopédie méthodique_. Each term defined in the dictionary was classified by the part of medicine it belonged in. Most of the articles on environmental topics were classed as hygiene. The entries describing air, airs, waters and places, Africa, climate, medical topography, topographical laws, and the tropics, were all so classed. (19) The definition given of hygiene itself by Hallé (1754-1822) in 1798 makes the point more explicitly. (20) A member of the Paris faculty in the _ancien régime_, he was made Professor of Medical Physics and Hygiene after the reorganisation of medical education in 1795. (21) In fact, in the 1840s, a prominent medical writer on hygiene gave him the credit for the establishment of hygiene as a separate field of medicine and for emphasising the importance of 'public hygiene'. (22) Hallé's article culminated in a schematic plan of the subject toform the framework of a treatise on hygiene. It was originally published in Fourcroy's journal, _La médecine éclairée par les sciences physiques_, (23) the very title of which reinforces one of the leading themes of this essay, the late eighteenth-century conviction that the physical and the life sciences were working together in a common endeavour. Hallé's projected treatise comprised four parts: an introduction on the natural history of mankind, part one on the subject of hygiene, part two on the substance of hygiene, and part three on the rules of hygiene. The introduction contained two sections, one geographical, the other historical. Medical geography, which revealed climatically caused variation in the human race, was complemented by the natural history of mankind, with an emphasis on temporal change. According to Hallé, the subject of hygiene, _sujet d'hygiène_, had two facets, the social and the individual. The former entailed studying the effect of climate, location, occupation, customs, laws, and government; the latter investigated how the variables of age, sex, temperament, habits, profession, poverty, and travel impinged on human health. The distinction between public and private was a difficult one to draw clearly. (24) The second part of Hallé's projected work dealt with _matière d'hygiène_, the substance of hygiene. His debt to Hippocratic, and particularly to Galenic, traditions is evident here. This part dealt with 'the things which mankind uses or handles, improperly called non-natural, and their influence on our constitution and organs'. (25) Hallé's use of '_improprement_', an adjective increasingly appended to the expression 'non-naturals' at the end of the eighteenth and the beginning of the nineteenth centuries, in no way lessens his debt to traditions of ancient medicine. It refers to his belief that the 'non-naturals' were in fact perfectly 'natural', that is, they were part of the regular order of nature, amenable to scientific analysis and were in fact those variables which should be looked to for providing law-like insights into the organism-environment relationship. It was this harnessing of the non-naturals to the 'airs, waters and places' approach of Hippocratic medicine which engendered a shift in the interpretation of this piece of Galenic medicine during the eighteenth century. (26) Hallé used Latin terms for the non-naturals, employing a vocabulary that may be original to him. He did however acknowledge a vague debt to 'the ancients' via Boerhaave's Institutes of medicine. This part of Hallé's plan is a veritable taxonomy, each non-natural was a class, while its subdivisions were orders. These orders in turn were sometimes subdivided. The six categories were, 'circumfusa', the physical environment; 'applicata', things applied to the surface of the body like clothes or soap; 'ingesta', substances introduced into the body, i.e. food and drink; 'excreta'; 'gesta', voluntary movements; and finally, 'percepta', functions of the nervous system, what had previously been termed the passions of the soul.

For our present concerns, the class 'circumfusa' is the most interesting. It comprised two orders, the atmosphere, and 'the earth, or places and waters', a category closely related to the famous Hippocratic treatise which inspired so much work in the area of hygiene. (27) Under 'atmosphere', Hallé included a number of agents commonly thought to impinge on the organism: air and its constituents, heat and light, electricity, magnetism, larger scale atmospheric changes associated with the seasons, and 'meteors' which included a wide range of atmospheric phenomena. The theme of the medical importance of air was echoed elsewhere in the Encyclopédie méthodique. The meteorologist, Cotte, wrote on the air and the atmosphere, analysing in turn each of its aspects, weight, humidity, heat, and so on, for their effect on the human body. Fourcroy, (1755-1809), contributed companion articles on the use of air as therapy, and on air as the cause of disease. (28) Extensive cross referencing led the reader on to the more detailed considerations of the atmospheric causes of disease in emanations, miasmas and mephitic substances.

The second order, that of airs, waters and places, was more geographical. It included climate, soil, 'natural changes' on the earth such as earthquakes and floods, and also the artificial changes in the environment resulting from the location of human settlements and the general effects of human society. Hallé's category, 'circumfusa', thereby summarises the prevalent contemporary conception of the sciences of the environment and their relevance to medicine. This included the study of geology, meteorology and both physical and social geography. Hallé's scheme also opened the way for astronomical changes to be taken into account, a theme made more explicit by Lamarck who considered the phases of the moon an important factor in biology. (29) Other writers concurred with this view, and emphasised the need to gather astronomical data for medical purposes:

> Following the father of medicine [i.e. Hippocrates] astronomical knowledge is absolutely vital to the doctor. The physician should observe exactly the rising and setting of the heavenly bodies, the equinoxes, the solstices and so on, the path of the sun, that of the moon and its phases, the season of the dog star, and of the seven stars, etc. (30)

Nor was this approach confined to France; in 1801 the New York-based Medical Repository published an unsigned article on 'Observations on the influence of the moon on climate and the animal economy; with a proper method of treating diseases when under the power of that luminary'. (31)

Hallé had also reiterated a common viewpoint in referring to electricity as an atmospheric agent of dramatic potency. As Bertholon (1742-1800) said in 1780: 'This very active and penetrating fluid, cannot exist in the atmosphere which surrounds us, in which we are submerged like a fish in water, without communicating itself to us, without affecting and influencing our bodies.' (32) The study of medical electricity went back several decades. One of its leading exponents by the end of the century, Mauduyt (1730?-1792), had contributed the article 'electricity' to the medical section of the Encyclopédie méthodique in 1792. (33) His article listed no fewer than twelve different medical applications of electricity for therapeutic purposes, a classic example of the management and control of natural agents for medical ends. His bibliography cited thirty-nine authors, French, German, Scottish, and English, who had contributed major works to the field. Medical electricity is a particularly good case of the synthesis, both practical and theoretical, of the physical and the biological sciences where an atmospheric agent was not only shown to affect human health in dramatic ways but was also harnessed by the medical profession for therapeutic purposes using elaborate instrumentation developed by natural philosophers and craftsmen. (34)

This transition, from isolating and understanding natural forces to using them for broader purposes, was well exemplified by Hallé's third and final part of hygiene, the discovery of 'rules for the preservation of mankind by the well regulated use of the things called non-natural'. Here the idea of management came to the fore, particularly in relation to individual or private hygiene. The discussion was cast in terms of régime, or regimen in English, a category which included far more than diet, although this was certainly a major consideration. It is important to remember that the words contained a range of meanings from the medical one implying the regulation of diet, exercise, in fact of mode of living in general, to the more general ones of the act of governing, a particular form of government, or a prevailing system. (35) Under the head 'General principles of regimen', Hallé listed four sets of parameters: use and abuse, excess and privation, regularity and irregularity, and habits and change. The theme of self-moderation pervades the hygiene literature of the period. Hallé, for example, also wrote on '*abstême*' and '*abstinence*' which discussed such matters as the control of alcoholic drinks. (36) Similarly, Johann Peter Frank (1745-1821) laid great stress on 'laws of moderation' and made it plain that society, more than the individual was at risk:

> The public intemperance, the excesses of an entire nation, and the shortcomings of unhealthy clothes that cripple the bodies, all these are not anymore the mistakes of an individual citizen; their consequences, therefore, cannot be damped with weak measures. They require the stronger hand of the official physician who, though he may permit a member of society to drown himself in wine in the dark, nevertheless cannot suffer a whole people to lose all its natural good qualities by debauchery. (37)

At the very end of his article in the *Encyclopédie méthodique*, Hallé added another section to clarify the relationship of hygiene to the actual practice of the healing art. The physician evidently had to understand how groups and individuals become predisposed to illness. Both epidemic and endemic disease played a part, as did the constitution of the individual patient. Doctors had to comprehend the systematic effects of the non-naturals and the accidental, unpredictable environmental causes of disease. However, the medical issue was not just knowing the etiology of environmentally caused illnesses, but of discovering the rules that preserve a positive state of health and of curing existing illnesses of all kinds.

Hallé made an appeal to the experimental physical sciences to become part of the medical enterprise, an invitation made concrete by the *Société Royale de Médecine*. The Society gathered extensive data from French provincial physicians on the relations between climate, geography and disease, and they asked correspondents to provide meteorological data taken three times a day. These elaborate findings were collated annually by Père Louis Cotte. Empirical precision, achieved through accurate, identical instruments was stressed by the Society. Collecting data on epidemics, i.e. common diseases of either animals or humans, was only the first step, for 'the Society attempted to suggest means by which possible agents might be removed'. (38) Lamarck's meteorological annuals (1799-1810) which have been much misunderstood must be located in this tradition. Like the project of the Society, and possibly even inspired by it, Lamarck's annuals idealised cooperative, accurate data collection, stressing those environmental factors which affect plants, animals and humans, and which are therefore of immediate practical utility to physicians and agriculturalists. (39) Lamarck's emphasis was more astronomical than that of some others, and elsewhere I have argued that this may account for the poor reception of his meteorological work at a time when the literature on environmental influences on life was extensive and growing. (40)

Geography and disease

Another example of the use of skills developed in conjunction with the physical sciences becoming basic to medical perceptions is the development of medical cartography. During the eighteenth century this took two forms: topographic maps displaying those features which the Hippocratic tradition recognised as relevant to health, such as marshes and prevailing winds, and maps which actually plotted the distribution of disease and which are later productions associated with epidemics of specific diseases, principally yellow fever and cholera. (41) Hallé's work on Africa relied heavily on the maps of Philippe Bouache (1700-1773) for the geographical data from which he deduced medical implications. (42)

There is no doubt that epidemics provided a spur to empirical investigation in the late eighteenth century. We must assume, however, that this is not because there were more of them then or because more people were killed but because ways of responding to epidemics changed. In particular, such bouts of illness and associated deaths could be treated quantitatively, using statistical methods. In fact, these phenomena were among a small number of biological phenomena which appeared ripe for the use of mathematical techniques which would reveal the specific environmental causes of epidemics.

Prevalent illnesses were also described in a fair amount of clinical detail and, respiratory diseases especially, they were linked to environmental causes, mostly to the air. Analysis of the air had been aided by work on the chemistry of gases. Imponderable fluids in the atmosphere were also thought to be fruitful explanations of the effects of different airs on health. In addition to work on perspiration and its effects, diseases were thought to be transmitted by emanations, miasmas,mephitic substances, even smells. As Desmarest (1725-1815) so charmingly put it when he wrote about the effect of atmosphere on health, exhalations from animals grazing on unusually fertile ground gave off such a potent smell that in summer the air hung heavy over the Swiss village he was describing. As a consequence, the inhabitants were small, ugly, and prone to cretinism. (43) Volney made a similar use of characteristics of the air to explain especially American illnesses such as catarrh and influenza, further suggesting that poisoned vapours caused epidemics but that they were so 'subtle' as to be undetectable. (44) The interest in yellow fever in late eighteenth-century America is a good example of the desire to understand the nature and distribution of a specific disease. Medical concern peaked following the big outbreak in Philadelphia in 1793. Unlike many problems of 'hygiene', yellow fever attacked rich and poor alike, and was particularly severe in whites. Consequently, it has been argued, it attracted far more medical attention than diseases confined to the poor and blacks such as scurvey, scrofula and scabies. (45)

A particularly rich source for investigating the medical community's views of environmental medicine is the journal which began publication in 1797, the _Medical Repository_. It was edited by Samuel Mitchill (1764-1831), physician and later Congressional Representative for the city of New York during Jefferson's presidency, and Edward Miller (1760-1812), also medically qualified. (46) In fact, the journal was largely founded, in the same spirit as the project of the _Société Royale de Médecine_, to collect data about the environment from a range of geographical locations for medical purposes. They were conscious of having a unique case study on their doorstep:

> In our extensive territory; in the varied descent, population, intermixture, institutions, manners and consequent diseases, of its inhabitants; in the opportunities it affords of observing and estimating the effects of old and new settlements, of gradual and rapid changes in the face of the country, of agriculture, commerce and navigation, of the savage, civilized, and intermediate states of society; of comparing the diseases, or phenomena of each disease, and the operation of the same remedies, in the same or different complaints, in Europe and America . . . (47)

The editors therefore requested that their readers furnish them with regular information on local diseases, local customs and diet, complete medical histories of their patients together with numbers of immigrants and when they arrived, environmental changes such as 'clearings, drainings' and increases in population. Similar data on domestic animals, on insects, especially pests, on vegetation, and on the state of the atmosphere were also solicited. The editorial of the first volume exuded confidence as to the success of the project:

> No plan seems more happily calculated to mark and explain the

> influence of different states of society, occupations, institutions, manners, exposure, air, modes of living, etc. etc. on health and thus, indirectly, on morals, industry and happiness . . . (48)

Ultimately, the idea was to understand the whole society in its environmental context, and this explicitly included the realm of work, as the editors' concern to gather information on occupationally caused diseases suggests. Subsequent issues of the journal contained numerous meteorological articles from many parts of the world, news of dramatic climatic events, reviews of geographical and hygienist books, and short communications on geological, palaeontological, and natural philosophical phenomena. Despite its international contents, the *Medical Repository* took seriously its mission to further the understanding of medical phenomena specific to the United States. Early issues, for example, put side by side meteorological information about New York and the tables issued by the New York hospital and the dispensary which catalogued the patients admitted by disease and gave the eventual outcome of the illness; patients either died, remained in care, were relieved or cured, and a few 'eloped'. (49)

Environmentalism and the science of man

Geographically limited medical problems offered the hope that specific environmental factors could be clearly isolated as causally relevant or not as the case may be. Adding the geographical dimension also suggested the fruitfulness of a comparative method which would further clarify why some illnesses seemed endemic in certain parts of the world. It therefore appeared relevant to Americans who were concerned about their own relatively new environment to compare their climate and diseases with those of Europe and South America. However, the natural history of mankind could not be assumed to be constant precisely because of the belief in the power of the environment to affect physical and social characteristics. The dynamic interaction between organisms and their environment was continual and it was hard to isolate a clear cause and effect relationship. It is important to remember that interest in the physical characters and social organisation of peoples in geographically distant areas antedates the environmental medicine literature by some time, playing a major role at the beginning of the eighteenth century. (50) This anthropological dimension manifested itself in a number of ways, from descriptions of exotic peoples and their habitats to detailing the manners, customs and modes of government of nations closer to home. The study of the different human races found a logical place in geographies and travel books and provided raw material for comparing European civilisation with 'savage' ones, a topic of particular interest to those who wanted to argue that civilisation itself engendered disease. The perspective of the _Société des Observateurs de l'Homme_, founded in 1799, exemplifies such interests. It aimed to attract members from a wide variety of backgrounds and included among its members Lamarck, Cabanis, Fourcroy, and many other _idéologues_. Just as Lamarck advocated the study of the simple invertebrates, so members of the Society urged the methodological advantages of studying 'primitive' people. Ideally, the observer should describe the physical environment first because of its powerful influence, this should be followed by an account of the physical characteristics of a typical member of society. It would be wrong however to suggest that there was total agreement on the interpretation of environmental influence. Volney was not a total determinist and he believed instead in the interplay between human-cum-social forces and environmental ones. Degérando, at least in his early work, thought human nature was fundamentally the same in all places and that its development was guided by natural laws of which the environment was the most important variable. His programme to 'perfect' the 'savage' by creating in him new needs and desires relied heavily on environmental manipulation. (51)

Hallé's lengthy description of Africa, in the 1787 volume of the _Encyclopédie méthodique_, illustrates some of these themes. He attempted to provide a total geography of the area based on rivers and mountain chains, climatic and seasonal changes, the soil and crops. This was the background for the medical interest in indigenous and, more especially, in immigrant groups, i.e. Europeans and creoles, and in their capabilities to adapt to the new environment. Following the example of James Lind, on whom he drew heavily, Hallé tried to formulate general rules to guide those who ventured into the African continent. (52) To reinforce the comparative nature of the enterprise, Hallé wrote a companion article on Europe and he hoped eventually to extend his analysis to include America and Asia. (53)

The _idéologue_ Volney's account of the _Soil and climate of the United States_ displayed similar preoccupations. He had left France in 1795, full of disillusion, in the hope of finding a haven in the New World, only to find by 1798 that there were places on that continent where Frenchmen were no longer welcome. There was in his mind, as in that of many Americans also, a comparison to be made between the two countries, both rebels in search of a new order. (54) The understanding of environment, health and customs were all part of the enterprise of evaluating the state of the society. The potential insights to be gained from studying the adaptation of the human race to its environment in America were, as the _Medical Repository_ had pointed out, enormous. Two racial groups attracted Volney's attention, the Indians and the white settlers. As a prelude to his study of the human dimension of the New World, Volney compiled a volume of observations, made by himself and others, of the general form and internal structure of the country, with chapters on waterfalls, earthquakes and volcanoes, climate, winds, lunar and solar influences on the weather, and prevailing diseases. His examination of the soil and climate was a 'suitable foundation' on which to base his projected second volume on the demography, history, laws, languages and culture of the United States. Volney's project was no innocent descriptive one; he wished to compare recent political events in France and America, judge the effect of revolution on them both, and draw moral lessons as he had done from earlier books on his travels in Egypt and Syria. (55) Mankind's ability to adapt to the environment was one criterion used in evaluating other cultures.

Volney was particularly scathing about the inability of white Americans to adapt their habits, especially diet and clothing, to their new surroundings. In this he was following writings he found in the _Medical Repository_, one of his principal sources on medical matters. In reviewing a book in 1802, the journal made the argument that was so attractive to Volney: Americans must liberate themselves from the tyranny of British habits if they are to survive. (56) Excessive tea drinking and meat eating, as well as heavy woollen clothing, were considered particularly inappropriate. No doubt general hostility to the British, on the part of both the Americans and the French, added weight to the argument. Yet it remains significant that the adaptation of habits to new environments should have been perceived as such an important issue.

The social environment

In attempting to unravel the effects of such a multitude of possible environmental agents, bills of mortality were an important source of information. They were sometimes published in the _Medical Repository_, and used by Noah Webster in his book which was voluminously reviewed and summarised in that journal. (57) His _Brief history of epidemic and pestilential diseases_ was, its reviewer claimed, the 'first attempt to write a history of epidemic and pestilential diseases'. (58) Its aim was succinctly summarised and its position in the environmental medicine tradition indicated:

> The leading object of this work is to prove that epidemic and pestilential distempers originate from a deleterious agent or agents, operating through the medium of the atmosphere . . . This is precisely the epidemic constitution of Sydenham . . . (59)

Bills of mortality helped determine whether illness was linked with other phenomena, such as the seasons, earthquakes, eruptions, and so on. Webster claimed to have demonstrated such links and postulated electric fluid as the common cause. The plea for public interest in the topic was repeated, coupled with a rejection of medical mystification in favour of a secular, naturalised medicine. (60) The utility of such a work as Webster's was, in the reviewer's opinion, very great. In fact, 'the legislator, the civil magistrate, the political economist, the military and naval commander', not to mention 'the patriot', would all benefit from reading the book. (61)

The climatic factor in disease was, by the end of the eighteenth century, widely recognised. It posed a simple management problem which entailed a limited range of solutions; few were more drastic than relocating dwellings, draining dirty water or moving to another climate for health reasons. That social conditions themselves posed a threat to health was far more difficult to deal with. This belief may be seen, for example, in early disenchantment with the consequences of civilisation when contrasted with a state of nature (however theoretical and unobtainable that might be), a position associated with Rousseau and Hallé. The problem of social conditions and ill health, as Frank had made plain, was essentially one of poverty, where overcrowded housing, poor working conditions, inadequate diet, and so on, were closely interrelated. In the _Medical Repository_ the issue of class was frequently made explicit. In describing climatic conditions in Londonderry, Dr William Patterson divided diet into three categories, that of the upper, middle and lowest (sic) classes. The point is that it was part of a list of environmental factors: topography, trees, geology, climate, temperature of the soil, architecture, the layout of the town, and social factors, including diet, according to social position. (62) The same volume of the _Medical Repository_ noted that some diseases of the previous winter had been caused by the use of rotten food among the poor. (63)

Professionalism and management

Two features must be noted which were central to the argument that social conditions affect health. First, the fear of the poor urban classes was a well established facet of British and French social commentaries in the second half of the eighteenth century. Paris and London had acquired the reputation of dangerous, depraved cities. It was thus already part of the common currency of middle and upper class culture to note the association between overcrowding and poverty, and between moral and physical degradation. Secondly, the medically trained argued that they possessed special expertise for dealing with social problems. The process of professionalisation of medicine is clear in descriptions given by the medically trained of their own work. In using the expression 'professionalisation', I am not referring to organised bodies or salaries or education, but to the feeling of being a special group with rights over a particular area of human life. The quotation from Frank, cited earlier in the paper, would be an excellent example. (64) The trend towards a self-conscious professionalism as manifested by efficiency and expertise has been noted in other areas of French life during the eighteenth century. (65) In the case of doctors, this professionalism also involved the claim that they had a special interest in certain social problems which they were uniquely well qualified to remedy. When Guillotin in 1790 asked the _Assemblée nationale_ to create a _Comité de santé_ he wished to exclude those who were not doctors from sitting on the committee because they were men who 'without beings doctors, are involved with natural philosophy, amuse themselves by studying the natural sciences, but by and large their knowledge is imperfect, insufficient and undisciplined'. (66)

At the very least, physicians should be the advisors and helpers of government; as Gilbert said in 1808, epidemics were totally avoidable if the medical profession and the law acted properly together. (67) The idea of medical police was a logical extension of the role doctors envisaged themselves playing, where their special scientific knowledge was harnessed, often explicitly as in the case of Frank, for the good of the state. It was precisely the fact that medicine spanned the natural and the moral realms that fitted it for the role of increasing the happiness of society, for as a French hygienist said, 'who is better suited than the physician to fulfil such a function, he who has made a deep study of nature and of the laws of physical and moral organisation?' (68)

Increasingly, the doctor also had a specific institutional terrain in which to exercise his professional power, which made him a doctor of society as well as of individuals. It has been pointed out that many of the institutions founded in the late eighteenth and early nineteenth centuries were inspired by environmentalist considerations. The total institution was a perfectly controlled environment which would cure the inmates both morally and physically. It provided a concrete place where prevalent beliefs about organism-environment relations were acted upon. I suggest this has to be seen in terms of management where the scientifically trained doctor used his expertise, and his newly acquired social position, to direct and control the patients' behaviour. Powerful supporting evidence for this comes from studies of hospitals, asylums, almshouses, and so on, in this period, which argue that there was a fundamental shift in social values which gave rise to these institutions, with the suggestion that the economic changes may have been fundamental causes. (69) While detailed analysis has been done for only a few cases, the argument could be applied more generally. Environmental medicine, and more specifically hygiene, seem particularly apt for such analysis containing as they do the belief in the power of external agents to affect human beings and the naturalistic linking of physical and social phenomena.

Naturalism

Perhaps the dominant character of the synthesis of earth science and environmental medicine was its naturalism, which avoided supernatural explanations by staying firmly within the realm of observable nature. I wish to refer to more than the increasingly secular character of eighteenth-century natural philosophy. Naturalism has been defined as studying 'a nature which worked through laws, which was simple, economical and had no waste'. (70) Two concepts of cardinal importance may be subsumed under naturalism - natural law and gradation. Natural law guaranteed that the universe worked in an orderly manner discoverable to the human mind. The regularities of nature should, it was argued, be used to formulate rules of conduct based on nature which guaranteed the well-being of the individual and of society. The metaphors of order and regularity, commonly associated with the Newtonian achievement in uncovering simple yet fundamental laws of nature, everywhere pervade the hygiene literature, only this time they apply to the desirable qualities of human behaviour. Nature was seen as a coherent, interconnected whole, with the best scientific explanations reflecting that by using as few terms as possible for a wide range of phenomena. Nature, furthermore, was not full of totally distinct classes of objects, rather they graded into one another, with the less complex being closely related to the slightly more complex and so on. A gradually increasing degree of organisation in the animal kingdom, for example, accounted naturalistically for the mental capabilities of man and the higher animals. This led to the position, held by Lamarck and Cabanis, that the moral and the physical were bound together, both having their source in the organisation of the organism. (71) The special mental qualities of humanity having been naturalistically linked to physical characters such as a large brain and well developed nervous system, social formations could also be understood in the same terms. For not only were societies products of the biological and psychological needs of the human family but their particular characteristics could be explained naturalistically by the impact on them of specific environments. This was the significance of Montesquieu's famous assertion of the influence of climate on the human race. In book fourteen of _De l'esprit des lois_, entitled '_Des lois dans le rapport qu'elles ont avec la nature du climat_', he argued that the human body, national character, religion, forms of government, customs, capacity for work, sobriety, illness, and culture, were affected by climate. Scientific naturalism made the moulding of social conditions by external forces as plausible as alterations in physical form.

Such an analysis was predicated on the vulnerability of both the individual and society to external agents, a view sustained by philosophical and psychological traditions which contributed their own environmentalist perspective. Locke's idea that the mind was a _tabula rasa_ at birth, and his emphasis on learning through education and life experience reinforced the idea that human beings were a product of their surroundings. Lockean theory had opened the way for a multitude of theories based on the importance of sensation rather than innate ideas in the development of the mind, as well as providing popular support for emphasis on learning good habits by correctly managed early experience. It was widely assumed that human experience would eventually be understood empirically by reference to 'ideas', 'sensations', and 'impressions'. This tradition, especially influential in France via Condillac and the _idéologues_, located the principal seat of environmental impact in the sense organs and nervous system. (72)

When allied with the common belief in the effect of the mental and moral on the physical, the biological approach which studied the nervous system provided a ready mode of conceptualising the power of the environment on life. Within medical, physiological and natural history circles there was growing support for a theory of the organism as responsive to external stimuli. The reactivity of the nervous system was reinforced by Albrecht von Haller's work on irritability and sensibility, the biological properties which came to symbolise organic reactivity. The Galenic tradition of the non-naturals, which emphasised air as a major determinant of health and disease, fostered similar convictions about organismic dynamism in the eighteenth century. This approach to life was further supported by theories of matter which postulated imponderable or very subtle fluids; climatic impact could partly be explained by perspiration for example, while reactions to electricity made sense if the nervous fluid was similar in nature to electric fluid. Other substances, like miasmas or mephitic vapours, also present in the atmosphere, might well enter the skin or lungs, yet not be visible to the naked eye.

The revival of Hippocrates in revolutionary France, by Cabanis and others, further illustrates the way in which conceptual trends and social policies went hand in hand. According to the Hippocratic corpus certain diseases, both endemic and epidemic, were associated with particular climates and topographies. These regular associations allowed the physician to predict both the predominant characteristics of the inhabitants, and the diseases to which they were most susceptible. Illness was a dislocation between organism and environment and therefore appropriate therapy took account of the key constituents of the environment such as diet and climate, while the overall ideal, like that of hygienist doctors, was prevention rather than cure. (73)

Hygiene and control

Thus, the new sciences of life and those of the environment contained within them theories which opened up the possibility of manipulating and managing the relationship between human beings and their surroundings. The model for this came partly from the experimental control already well established in the physical sciences. But the ethic of management had other more complex origins, especially in political thought. The early eighteenth century for example saw the establishment of chairs of Polizeiwissenschaft, the science of police, for the instruction of civil servants in the techniques of management. (74) It was in this context that Frank developed the idea of medical police, that is, the administration of the health aspects of the state by those medically trained especially in public health. Management of the organism-environment bond was exemplified by the new meritocrats, like the idéologues in France, who were policy makers in areas where manipulation of that bond, through schools, hospitals etc, was thought desirable. Similarly, the environmentalist assumptions of early American penitentiaries, asylums and almshouses has been demonstrated. (75) The belief that deviance, being environmentally caused, could be eradicated in the same way was at the heart of early nineteenth-century debates about the efficacy of solitary confinement. Not only was the penitentiary 'a grand theatre, for the trial of all new plans in hygiene', but its aim was to teach 'the habits of order and regularity'. (76)

Having acknowledged that the environment was a major determinant of human characteristics, it seemed reasonable to control the variables to produce desirable human features. Among the prime movers in this were the medical profession, a growing and powerful body of middle class men who were intimately involved in the construction and maintenance of the social order. The links of environmental medicine to public policy seem clear enough. In the case of Frank, the cameralist tradition led to an emphasis on population as the source of wealth so that the measures he advocated were aimed at increasing the population and maintaining order. The hygienist emphasis on order, discipline, regularity, sobriety, in fact on moderation in all things, must be seriously examined. There was in the late eighteenth century extensive debate about these issues and about what measures would best secure social lawfulness.

A later writer in the environmental medicine tradition, Robley Dunglison, provides some hints of how to look at the social roots of hygiene. In discussing the medical problems of poverty he was by no means original - Frank had given a famous oration in 1790 on 'The people's misery - mother of diseases', and by the 1820s Villermé's detailed work was well under way. (77) In 1844 Dunglison opened his work by asserting that it was the domestic condition of the poorer classes rather than their industrial situation to which their ills must be ascribed. Dunglison shared contemporary interest in occupational illness but believed that stunted children derived their disability from the town they lived in and from their domestic situation rather than from the factory. He further stated that such deformity had no ill effects. (78) This is reminiscent of the Medical Repository stating in 1801 that although coal smoke was unpleasant, it was in fact more healthy than the wood fumes which were common previously. (79)

When looked at in this light, the traditional historiography of public health in terms of high minded philanthropy begins to look inadequate. Foucault has already warned us that we must begin to look beneath the surface and see what it concealed, both in the primary sources of the period, with their claims to reform individuals and society, and by implication in the secondary literature, which reproduces the simple 'humanitarianism' of such enterprises. (80) At the very least, one can say from examining the evidence in its historical context that environmental medicine served the interests of the dominant class, for example, in suggesting that social or industrial problems had purely individual or domestic causes, and hence solutions. The language and concerns of hygiene suggested the need for a well ordered, moderate life style, using metaphors of discipline, regulation and self-denial. This synthesis of the environmental and life sciences using the language of the late Enlightenment not only gained in status by its scientific credentials but was also moulded by the reality and needs of everyday life which shaped people's consciousness of issues such as these which spanned a range of human activities.

Conclusion

The fact that environmental medicine and the observational sciences of geology, meteorology and geography marched hand in hand may appear strange. Both these areas of science shared the characteristics of their time. It has been said that 'enlightenment culture was more confident and aggressive towards the physical environment than any previous'. My argument is that the 'appropriation' and 'exploitation' embodied in this stance was applied to the human part of nature as well as to its physical counterpart. (81)

It might be tempting to see the common synthesis of the environmental and bio-medical sciences in Germany, France, Britain, and the United States as indicating minimal cultural differences and suggesting the independence of science from its social context. I do not believe that such a conclusion is justified. Instead, it might be profitable to search for the common structural features of those societies which might underlie this constellation of beliefs and practices.

Recent work on a closely related problem in comparative history may prove helpful in suggesting avenues for further analysis of this kind. (82) The establishment of large custodial institutions to replace community based 'outdoor relief' in the late eighteenth and early nineteenth centuries, like the science of hygiene, has been seen in traditional historiography as a great step forward in medical and social policy. More recently, it has also been shown that explanations such as rapidly increasing urbanisation cannot satisfactorily account for the range of nations which instituted the programme. Instead, we might emphasise changes in social 'structures, perceptions and outlook' and the associated transition towards a new type of economy which took place in these countries at the end of the eighteenth century. Social class played a central role, for there was growing dissatisfaction on the part of the bourgeoisie who wished to instill their values on the section of the population which appeared least tractable, or in the language of hygiene, most unhealthy. (83) If institutions were one way to do this, I am suggesting that environmental medicine, and specifically hygiene, was another.

Without rehearsing the detailed arguments of works which have undertaken such analyses, perhaps we can draw a lesson from this approach. (84) The search should be for deep structural features which, unlike those described by Foucault, serve to link science and medicine with specific socio-economic formations which touched the lives and consciousness of the whole community. This emphasis on the human roots of seemingly theoretical concerns seems especially apt for the present case. The literature I have described which synthesised earth science and environmental medicine was rooted in concrete reality. We should therefore analyse it in those terms, and restore to the sciences of the environment the central role they played in this period in the control and mastery of the physical and social worlds.

Notes

The author wishes to thank Joyce Oldham Appleby, David Bien and Karl Figlio for their help and support. I am particularly grateful to Caroline Hannaway for supplying me with information about the *Société Royale de Médecine*.

This paper is dedicated to Joyce Oldham Appleby.

1 Anonymous writer, Medical Repository, [hereafter cited as M.R.], 1800, iii, 390.

2 The literature on the environmental sciences during the eighteenth century is sparse. On geology and the earth sciences see R. S. Porter, "The making of geology: earth science in Britain, 1660-1815", Cambridge, 1977; D. H. Hall, "History of the earth sciences during the scientific and industrial revolutions", Amsterdam, 1976. On meteorology see A. K. Khrgian, "Meteorology. A historical survey", Jerusalem, 1970; W. E. Knowles Middleton, "A history of the theories of rain and other forms of precipitation", London, 1965. P. C. Ritterbush, "Overtures to biology. The speculations of eighteenth-century naturalists", New Haven & London, 1964, contains a useful bibliography of less well known primary sources based on meteorological and electrical observations. Among the few secondary sources which deal with the synthesis I describe is M. Grmek, 'Géographie médicale et histoire des civilisations', Ann. Écon. Soc. Civilisations, 1963, xviii, 1071-97.

3 G. Stocking, 'French anthropology in 1800', Isis, 1964, lv, 134-50.

4 C. de Volney, "Voyage en Syrie et en Egypte", Paris, 1787; idem, "A view of the soil and climate of the United States of America" (ed. G. W. White), New York & London, 1968.

5 E. H. Ackerknecht, 'Hygiene in France, 1815-1845', Bull. Hist. Med., 1948, xxii, 117-55, contains a useful bibliography of primary sources. William Coleman, University of Wisconsin, Madison, is currently undertaking a detailed study of Villermé.

6 G. Rosen, "From medical police to social medicine. Essays on the history of health care", New York, 1974, pp.120-58.

7 K. M. Figlio, 'The metaphor of organisation: an historiographical perspective on the bio-medical sciences of the early nineteenth century', Hist. Sci., 1976, xiv, 17-53.

8 C. L. Glacken, "Traces on the Rhodian shore. Nature and culture in western thought from ancient times to the end of the eighteenth century", Berkeley, 1967.

9 C. Hannaway, 'The Société Royale de Médecine and epidemics in the ancient régime', Bull. Hist. Med., 1972, xlvi, 257-73; Glacken, op.cit. (8), pp.659-61, 669-71, 697.

10 Glacken, op.cit. (8), vii-viii.

11 R. Hooykaas, 'The parallel between the history of the earth and the history of the animal world', Arch. Int. Hist. Sci., 1957, x, 3-18.

12 Volney is a good example of this trend as is P. C. Levesque (1725-1815), well known for his histories of France, Russia and Ancient Rome. As a member of the second class of the Institut, he wrote on the development of the human race and on definitions of nature. See for example his 'Considérations sur l'homme, observé dans la vie sauvage, dans la vie pastorale, et dans la vie policé', Mémoires de l'Institut national des sciences et

arts . . . sciences morales et politiques, 1796, i, 209-46. Others in the geography section of the second class continued to work in the traditions of P. S. Pallas (1741-1811) and N. Desmarest.

13 Glacken, op.cit. (8), pp.565-81: O. H. K. Spate, 'Environmentalism', "International encyclopedia of the social sciences", (ed. D. L. Sills), 1968, v, 93.

14 Villermé, for example, had an enormous influence on nineteenth-century writers, including Jules Michelet, who used him as an authority on the ill effects of poverty and destitution on women; see, for example, L. J. Jordanova, 'Women and the social order: the views of Jules Michelet', 1978, unpublished paper; see also note (5).

15 H. Baudet, "Paradise on earth: some thoughts on european images of non-european man", New Haven, 1965; J. N. Hallé, 'Afrique', "Encyclopédie méthodique, médecine" (hereafter cited as E.M.M.), 1787, i, 281-353; Stocking, op.cit. (3).

16 F. Picavet, "Les idéologues. Essai sur l'histoire des idées et des théories scientifiques, philosophiques, religieuses, etc. en France depuis 1789", Paris, 1891; H. F. May, "The Enlightenment in America", New York, 1976; Hannaway, op.cit. (9); Ackerknecht, op.cit. (5); Rosen, op.cit. (6).

17 E. Littré, "Dictionnaire de la langue française", 4 vols & supp., Paris, 1878, ii, 2073. On the general theme of hygiene see the special issue, 'Le sain et le malsain', Dix-huit. Siècle, 1977, ix.

18 R. Dunglison, "Human health: or the influence of atmosphere and locality; change of air and climate; seasons; food; clothing; bathing and mineral springs; exercise; sleep; corporeal and intellectual persuits, etc. etc. on healthy man; constituting elements of hygiene", new edn., Philadelphia, 1844, pp.iii-iv; "Oxford English dictionary", 1933, v, 493.

19 Wherever possible, brief biographical details are given of less well known contributors to the Encyclopédie méthodique. J. B. Huzard, 'Air', E.M.M., 1787, i, 488-92; unsigned, 'Air, eaux et lieux', E.M.M., 1787, i, 590; Hallé (member of the Académie Royale des Sciences and associé ordinaire of the Société Royale de Médecine), op.cit. (15); L. C. R. Macquart (1745-1818 , associé ordinaire of the Société Royale de Médecine), 'Climat', E.M.M., 1792, iv, 878-9; Thillaye ainé, 'Topographie médicale', E.M.M., 1830, xiii, 278; J. J. De Brieude (1729-1812, Royal Physician, associé ordinaire of the Société Royale de Médecine), 'Lois topographiques', E.M.M., 1808, viii, 185-93; Thillaye ainé, 'Tropiques', E.M.M., 1830 , xiii, 322.

20 J. N. Hallé, 'Hygiène', E.M.M., 1798, vii, 373-437.

21 D. M. Vess, "Medical revolution in France 1789-1796", Gainesville, 1976.

22 Dunglison, op.cit. (18), p.iii; 'since the time of M. Hallé more especially, public and private Hygiene have formed part of the curriculum of study at the École de Médecine of Paris'.

23 Hallé, op.cit. (20), pp.433-7. Fourcroy's journal appeared fortnightly in Paris in 1791 and 1792. Like Hallé, Fourcroy was also a member of the Académie Royale des Sciences and the Société Royale de Médecine.

24 Dunglison, op.cit. (18), p.iii.

25 Hallé, op.cit. (20), p.433; (my translation; all quotations from the French have been translated by me unless the original citation is to a translation). The expression 'non-natural' may require clarification. According to Temkin, in Galenic thought 'medicine has two parts, theoretical and practical; the theoretical consists in the study of things natural . . . non-natural (hygiene), and contra-natural (pathology)'. The non-naturals were the cause of good health and could be used to prevent disease. The term itself originated with the medieval Galenists and the concept remained influential until the nineteenth century. O. Temkin, "Galenism. Rise and decline of a medical philosophy", Ithaca & London, 1973, pp.106, 102, 180.

26 L. J. Rather, 'The "six things non-natural": a note on the origins and fate of a doctrine and a phrase', Clio Med., 1968, iii, 333-47; P. H. Niebyl, 'The non-naturals', Bull. Hist. Med., 1971, ilv, 486-92; O. Temkin, op.cit. (25); W. Coleman, 'Health and hygiene in the Encyclopédie: A medical doctrine for the bourgeoisie', J. Hist. Med., 1974, xxix, 399-421.

27 P. J. G. Cabanis, "Oeuvres philosophiques" (ed. C. Lehec and J. Cazeneuve), 2 vols., Paris, 1956; A. Coray, "Hippocrates. Traité des airs, des eaux et des lieux", 2 vols., Paris, 1800; G. Miller, '"Airs, waters and places" in history', J. Hist. Med., 1962, xvii, 129-40; B. Aschner, 'Neo-Hippocratism in everyday practice', Bull. Hist. Med., 1941, x, 260-71.

28 L. Cotte, 'Air et atmosphère', E.M.M., 1787, i, 579-90; A. Fourcroy, 'Air atmosphérique (materia medica)'; 'Air atmosphérique (pathologie); 'Air vital ou déphlogistique', E.M.M., 1787, i, 575-6, 577, 578; see also W. A. Smeaton, 'Fourcroy', "Dictionary of scientific biography", 1972, v, 89-93.

29 J. B. Lamarck, "Annuaires météorologiques", 11 vols., Paris, 1799-1810; L. J. Jordanova, 'The natural philosophy of Lamarck in its historical context', University of Cambridge PhD thesis, 1976, pp.79-84; on Lamarck see also, R. W. Burkhardt, "The spirit of system. Lamarck and evolutionary biology", Cambridge, Mass., 1977.

30 De Brieude, op.cit. (19), p.187.

31 M. R., 1801, iv, 285.

32 P. Bertholon, "De l'éléctricité du corps humain dans l'état de santé et de maladie" , Lyon, 1780, p.xiv.

33 P. J.C. Mauduyt de la Varenne (associé ordinaire of the Société Royale de Médecine), 'Électricité', E.M.M., 1792, v, 709-49; see also his "Mémoire sur les différentes manières d'administrer l'éléctricité et observations sur les effets qu'elles ont produit", Paris, 1792.

34 The work of George Adams the younger (1750-1795), an instrument maker, is worth noting. Among his many publications on natural philosophy which centred on the use of instruments, see "An essay on electricity, explaining the theory and practice of that useful science; and the mode of applying it to medical purposes", 2nd edn., London, 1785. Another case of the harnessing of electrical and magnetic forces through instruments for medical purposes was mesmerism: see R. Darnton, "Mesmerism and the end of the Enlightenment in France", Cambridge, Mass., 1968; F. Rausky, "Mesmer ou la révolution thérapeutique", Paris, 1977.

35 Littré, op.cit. (17), iv, 1558-9; "Oxford English dictionary", 1933, viii, 369.

36 J. N. Hallé, 'Abstême', 'Abstinence', E.M.M., 1787, i, 35-6, 37-8.

37 J. P. Frank, "A complete system of medical police" (ed. with an introduction by E. Lesky), Baltimore, 1976, p.154.

38 Hannaway, op.cit. (9), 269; J. P. Peter, 'Malades et maladies à la fin du XVIIIe siècle', Ann. Écon. Soc. Civilisations, 1967, xxii, 711-51.

39 Lamarck, op.cit. (29).

40 Jordanova, op.cit. (29), pp.82-3.

41 L. G. Stevenson, 'Putting disease on the map, the early use of spot maps in yellow fever', J. Hist. Med., 1965, xx, 227-61; S. Jarcho, 'Yellow fever, cholera and the beginnings of medical cartography', J. Hist. Med., 1970, xxv, 131-42.

42 On Bouache see N. Broc, 'Un géographe dans son siècle: Philippe Bouache (1700-1773)', Dix-huit. Siècle, 1971, iii,223-35; also relevant are N. Broc, 'Voyages et géographie au XVIIIe siècle', Rev. Hist. Sci., 1969, xxii, 137-54, and S. Moravia, 'Philosophie et géographie à la fin du XVIIIe siècle', Stud. Voltaire, 1967, lvii, 937-1021.

43 N. Desmarest, 'Atmosphère. Ses modifications relatives à la salubrité des habitans', "Encyclopédie méthodique géographie physique", 1803, ii, 841.

44 C. de Volney, "A view of the soil and climate of the United States of America" (ed. G. W. White), New York & London, 1968, pp. 234-7.

45 R. H. Shryock, "Medicine and society in America: 1660-1860", New York, 1960, p.94.

46 There was occasionally a variation in editorship in the early years, but Mitchell and Miller seem to have been the stable foundation of the enterprise. More information about them may be found in Shryock, op.cit. (45); and in H. F. May, "The Enlightenment in America", New York, 1976; see also C. R. Hall, "A scientist in the early republic: Samual Latham Mitchill, 1764-1831", New York, 1934; S. Miller, "Medical Works of Edward Miller", New York, 1814.

47 M.R., 1797, i, viii.

48 M.R., 1797, i, x.

49 M.R., 1797, i, 105, 245-9, 373-9, 557-64.
50 Baudet, op.cit. (15), pp.39-42.
51 Stocking, op.cit. (3).
52 J. Lind, "A treatise on the scurvey", Edinburgh, 1753; "Three letters relating to the navy, Gibraltar and Portmahon", London, 1757; "An essay on diseases incidental to Europeans in hot climates with the method of preventing their fatal consequences. To which is added an appendix concerning intermittant fevers. To the whole is annexed a simple and easy way to render salt water fresh, and to prevent a scarcity of provisions in long voyages at sea", London, 1768. See also Jordanova, op.cit. (29), pp.70-73.
53 J. N. Hallé, 'Europe', E.M.M., vi, 200-41.
54 J. Appleby, 'America as a model for the radical French reformers of 1789', William Mary Quart., 1971, third series, xxviii, 267-86.
55 Volney's own testimony supports this point, see Volney, op.cit. (44), pp.v-xxv. See also May, op.cit. (46), pp.225, 231, 234, 246.
56 M.R., 1802, v, 62-6. The occasion for the remarks was the review of a book by Charles Caldwell, "Medical and physical memoirs, containing, among other subjects, a particular inquiry into the origin and nature of the late pestilential epidemics of the United States", Philadelphia, 1801.
57 N. Webster, "A brief history of epidemic and pestilential diseases; with the principal phenomena of the physical world, which precede and accompany them, and observations deduced from the facts stated", Hartford, 1799. For the review see M.R., 1800, iii, 278-88, 390-7. For more information on Webster's role in the American Enlightenment see May, op.cit. (46), pp. 192-3, 236. Webster was a frequent contributor to the Medical Repository in its early years, usually in the form of a letter communicating his ideas to Dr Miller. For example, 'On the connection of earthquakes with epidemic diseases, and on the succession of epidemics: communicated in a letter from Noah Webster, to Dr Miller, dated New-Haven February 21 1801', M.R., iv, 340-4; 'A collection of phenomena, relative to the connection between earthquakes, tempests, and epidemic distempers; and a vindication of the doctrine of equivocal generation', M.R., 1802, v, 25-31.
58 M.R., 1800, iii, 279.
59 Ibid.; I. Galdston, 'The epidemic constitution in historic perspective', Bull. N.Y. Acad. Med., 1942, viii, 606-19.
60 M.R., 1800, iii, 288.
61 Ibid.
62 W. Patterson, 'Observations on meteorological instruments and on the weather at Londonderry in the year 1797', M.R., 1800, iii, 36-46.

63 M.R., 1800, iii, 63; this information was given under the head, 'Medical and philosophical news'.

64 See note (37). There is a growing literature on the history and sociology of the medical profession: I. Waddington, 'The struggle to reform the Royal College of Physicians, 1767-1771: a sociological analysis', Med. Hist., 1973, xvii, 107-26, and 'General practitioners and consultants in early nineteenth-century England: the sociology of an intra-professional conflict', in J. Woodward & D. Richards (eds.), "Health care and popular medicine in nineteenth century England", London, 1977; N. Parry & J. Parry, "The rise of the medical profession. A study in collective social mobility", London, 1976; J. L. Berlant, "Profession and monopoly. A study of medicine in the United States and Great Britain", Berkeley, 1975.

65 D. Bien, 'Military aristocrats in the French Enlightenment: toward counter-revolution', unpublished paper, cited by permission of the author; for his treatment of the army and its officers see also 'La réaction aristocratique avant 1789: l'example de l'armée', Ann. Écon. Soc. Civilisations, 1974, xxix, 23-48, 505-34. Bien's general thesis is that, during the eighteenth century, there was an increasing interest in having a specific profession rather than in having a rank. He has suggested that 'profession or function appears to have been moving to the center of the individual's sense of self' ('Military aristocrats', p.4).

66 Cited in D. B. Weiner, 'Le droit de l'homme a la santé - une belle idée devant l'assemblé constituente', Clio Med., 1970, v, 209-23 (212).

67 Gilbert, 'Maladies épidémiques (police médicale)', E.M.M., 1808, viii, 376-82.

68 Ackerknecht, op.cit. (5), p.124.

69 D. J. Rothman, "The discovery of the asylum. Social order and disorder in the new republic", Boston, 1971; A. T. Scull, "Decarceration. Community treatment and the deviant - a radical view", Englewood Cliffs, N.J., 1977; M. Foucault, "The birth of the clinic", London, 1973.

70 R. S. Porter, 'The physical environment', in R. S. Porter & G. S. Rousseau (eds.), "The ferment of knowledge: changing perspectives in scholarship of eighteenth-century science", Cambridge, forthcoming.

71 J.B. Lamarck, "Philosophie zoologique", 2 vols., Paris, 1809, i, 364; Figlio, op.cit. (7).

72 Picavet, op.cit. (16), is good on the seventeenth-century origins of <u>idéologie</u>; see also, I. F. Knight, "The geometric spirit. The Abbé Condillac and the French Enlightenment", New Haven, 1968; K. M. Figlio, 'Theories of perception and the physiology of mind in the late eighteenth century', Hist. Sci., 1975, xii, 177-212.

73 For a detailed analysis of Cabanis and his medical philosophy see

M. S. Staum, 'Cabanis and the science of man', Cornell University PhD thesis, 1971.

74 Frank, op.cit. (37), p.xv.

75 Rothman, op.cit. (69), pp.82, 129, 193, 238, 245.

76 Ibid., p.83.

77 J. P. Frank, 'Akademische Rede vom Volkselend als der Mutter der Krankheiten (Pavia 1790)', Sudhoffs Klassiker der Medizin, 1964, xxxiv. L. R. Villermé, "De la famine et de ses effets sur la santé dans les lieux qui sont le théâtre de la guerre", Paris, 1818; "Des prisons, telles qu'elles sont et telles qu'elles devraient être par rapport à l'hygiène, à la morale et à l'économie", Paris, 1820. Villermé was a prolific author on issues in social and environmental medicine and on topics in hygiene; these are among his earliest publications. See also Ackerknecht, op.cit. (5), p.153 for more bibliography, and also M. Perrot, "Enquêtes sur la condition ouvrière en France au 19e siècle", Paris, 1972.

78 Dunglison, op.cit. (18), pp.i, 53.

79 M.R., 1801, iv, 258.

80 Foucault, op.cit. (69); see also his earlier works "Madness and civilization", New York, 1965, and "Mental illness and psychology", New York, 1976.

81 Porter, op.cit. (70).

82 Scull, op.cit. (69); see also Rothman, op.cit. (69) and Foucault, op.cit. (69) and (80).

83 Scull, op.cit. (69), pp.25-6.

84 See, for example, K. M. Figlio, 'Chlorosis and chronic disease in nineteenth-century Britain: the social constitution of somatic illness in a capitalist society', Social Hist., 1978, iii, 167-97.

Chemical geology or geological chemistry?

W. H. BROCK

The 1850s and 1860s were significant decades not only for the fragmentation of the sciences into specialised disciplines, each with their own methods, techniques, approaches and subject boundaries, but also for the budding, branching and institutionalisation of 'interdisciplinary' sciences, of which astrochemistry (intriguingly rapidly relabelled astrophysics) and biochemistry are the most familiar examples.(1) Echoing Spencer's edict that 'the sciences are arts to each other', (2) David Allen has pointed out that individual sciences have natural affinities that form a 'mental cartography . . . which preconditions the extent to which a discipline receives influences from its neighbours'. (3) Because chemistry was never one of the field sciences that made up the cosy family unit of natural history, its relationship with open-air geology was historically much less intimate than that between geology, botany and zoology. However, as Weindling says, (4) if geology formed the words of the book of nature, and mineralogy its alphabet, chemistry was its potential grammar. Botany and zoology could aid the stratigrapher and chronologist; mathematics and physics could help solve the puzzling nature of the earth's interior; mineralogy could identify the materials in the earth's crust; optics and microscopy might extend the geologist's vision of the physical nature of rocks; but to go beyond physical appearance to explanations of the earth's history - to questions concerning the likely modes of formation and origin of rocks - demanded chemical knowledge. (5)

Because of chemistry's growing utility to the mineralogist and miner, chemistry and geology were increasingly forced into one another's company from the mid-eighteenth century onwards: Lavoisier probably gravitated towards chemistry from the inspiration of geological problems, Berzelius developed a chemical, as opposed to physical classification of minerals, and numerous chemists, including Black, Davy and Thomson found themselves obliged to lecture on geology. (6) By the 1830s, chemists were being regularly consulted by geologists to analyse ore and soil samples. Moreover, through the agencies of the British Association for the Advancement of Science, the Royal College of Chemistry, and the Royal Agricultural Society, the long-standing tradition of landed interest in agricultural and estate improvements further promoted 'the applications of chemistry, particularly of analytical chemistry, to the problems of landownership'. (7) It was this interest and the economic significance of ancillary chemistry which enabled Henry De La Beche to establish a permanent chemist's post within the Mining Records Office of the Museum of Economic Geology in 1839 - a position which was soon paralleled in the several North-American Geological Surveys that began their labours during the same decades. (8)

Thus at the same time that its sisterhood with the natural history sciences was becoming looser, geology was forming a new relationship with chemistry. In the mid-Victorian period both chemistry and geology were professional activities to an extent that neither botany and zoology were until the end of the century; moreover, as geology aspired more and more to exact measurement under the stern paternal eye of temporal physics (9) and became, consequently, more of a laboratory science, a marriage (rather than chaste sisterhood) with chemistry - the doyen of laboratory disciplines - seemed at the very least worth exploring. But if a marriage was to be arranged, who should and would be the dominant partner? Should geologists approach their rocks, mountains, stratifications, mineral veins, trapped waters, and volcanoes as chemists, or as geologists equipped with chemical tools, and educated in chemical concepts? Should they be chemical geologists or geological chemists?

These questions formed the core of an interesting debate which took place in the 1860s between the Manx geologist David Forbes (1828-76) and the North-American chemist and geologist Thomas Sterry Hunt (1826-92). Both men agreed that geology and chemistry were marriage partners, but whereas Forbes saw geology as the dominant partner, Hunt took the opposite view. Their debate, which attracted considerable attention in both the geological and chemical press, (10) not only explicitly defined the received relationship between geology and chemistry during the mid-Victorian period; it also had implications for the education of geologists, and raises questions for historians concerning the origin of the interdisciplinary science of geochemistry.

The protagonists

Whatever their respective contemporary reputations within the international geological community during the mid-nineteenth century, neither Forbes nor Hunt has found a niche in the received historiography of geology. Forbes has been overshadowed by Sorby, whom he admired; while Hunt ploughed a philosophical furrow that few laboratory-smitten chemists and geologists wished to follow. For these reasons, and because the formative influences on their careers helps to explain the content and tone of their controversy, we must begin with some biographical remarks. (11)

The younger brother of the witty Manx naturalist, Edward Forbes, David Forbes was educated at the University of Edinburgh. (12) In 1845 he spent some months in a metallurgical laboratory which another Edinburgh graduate, John Percy, had opened in Birmingham in order to supplement his appointment as physician at the city's Queen's Hospital. Percy, who was then developing a new method of extracting silver from its ores, and who was interested in geology, (13) brought Forbes to the attention of the nickel refiner and merchant, Brook Evans, who sent him to Norway in 1848. Here he was immediately made superintendent of the mining and metallurgical colony at Espedal, where he spent ten years, receiving Royal favours for arming miners in support of the King of Sweden in the revolutionary year of 1848. For his many contributions to, and knowledge of, Norwegian geology, Forbes was elected FRS in 1856. In 1857, prior to his resignation from Espedal, he visited Chile, Peru and Bolivia in search of exploitable nickel and cobalt ores. Following the failure of a proposal to appoint him a representative of the British government in South America to counter advances being made by German traders in that continent (a position for which he was widely supported), (14) Forbes made further extensive travels in Chile and in the South Sea Islands, whose volcanic origins and mineralogy he studied. Finally, in 1866, following further travels in Africa and Europe, he settled in London where, on the foundations of the Iron and Steel Institute in 1869, he became its conscientious Foreign Secretary. From 1871 until his death in 1876, he was also a Secretary of the Geological Society.

Whilst in Norway, and no doubt stimulated by his work with Percy, Forbes took advantage of the metallurgical operations he directed to submit various rocks to high temperatures and pressures for long or short periods, imitating (he claimed) metamorphic action in the production of various rock structures. (15) With his global experience of geology and metallurgy it is not surprising that on his return to England Forbes became one of the first to support Sorby's use of the microscope in the elucidation of geological phenomena, becoming Sorby's 'close friend, disciple and vigorous champion'. (16) Indeed, until Zirkel's publications in the 1860s demonstrated that Sorby's work could not be ignored, Forbes's support was exceptional.

Distinguished, as The Times noted, (17) as chemist, geologist and mineralogist, Forbes took especial interest in the bearings of chemistry on igneous and cosmic phenomena - studies which he felt were unduly neglected in Great Britain. (18) It was for this reason that the work of Sorby and the speculation of Hunt so excited him.

Two years older than Forbes, the origins, education and experience of Thomas Sterry Hunt were very different. (19) Born in Norwich, Connecticut, of Puritan stock, Hunt was virtually self-educated in science until be became a laboratory assistant to Benjamin Silliman Jr., the Professor of Chemistry at the University of Yale, for whom he wrote the organic chemistry section of a textbook. It was Silliman who found Hunt the job of analytical chemist to C. B. Adams' Geological Survey of Vermont in 1846, from which he graduated in 1847 to the important position of mineralogist and chemist to William Logan's Geological Survey of the then Province of Canada. (20) Whilst with the Survey, which was based in Ottawa, he was able to lecture (in French) at the University of Laval in Quebec, and on chemistry and mineralogy at the University of McGill in Montreal. Angry at not being appointed Logan's successor, Hunt left the Survey in 1872 in order to teach geology at the Massachusetts Institute of Technology in Boston, retiring in 1878 to follow a career of consultancy and lapsing into increasing obscurity. Hunt frequently visited France and England, where he was known to members of both the geological and chemical communities. Elected to the Royal Society in 1859, he was one of the organisers of the first International Congress of Geology in Paris in 1878, and was President of the Royal Society of Canada in 1884.

In collaborating with Logan on the latter's The geology of Canada (1863), Hunt was responsible for the sections on petrography, mineralogy, mineral waters and most of the chapters on economic geology. Logan and the Survey allowed Hunt complete freedom to develop distinctive geological ideas, especially of Paleozoic rocks and the older crystalline and metamorphic rocks of the earth's crust. Although he made several important observations in the field (including the relationship between petroleum deposits and anticlinal folds), (21) Hunt regarded himself primarily as a chemist, and always claimed to approach geology from a chemical standpoint. However, although Hunt's earliest papers had implied that he was an orthodox chemist in sympathy with the new unitary and type theories of organic chemical classification that were being established in Europe during the 1850s, his chemical philosophy was, in fact, idiosyncratic and heterodox. In the long term it generated a highly personal chemistry, geology and natural system of classification of minerals which, probably, no other contemporary scientist wished to share. (22)

By 1853, prompted by L. Oken's Elements of physiophilosophy (1847) and J. B. Stallo's General principles of the philosophy of nature (1848), Hunt abandoned atomism and juxtaposition of elements as explanations of chemical union for the Kantian doctrine of 'interpenetration of masses'. (23) Through Stallo, he agreed with Hegel that 'the chemical process is an identification of the different and a differentiation of the identical'. (24) Using what appears to be embryological analogies, he accordingly postulated only two kinds of chemical reaction: 'metagenesis', in which one chemical species differentiates into several others, or vice versa, where two or more species coalesce into one; and 'metamorphosis', in which a substance is sensibly transformed by condensation or expansion. (25) Stallo's and Oken's cosmology, together with the revived nebular hypothesis, (26) equipped Hunt with a developmental classification of the sciences in which their sequential order dictated their relative importance. An historic chemistry thereby became bonded to, and preceded an historical geology. Accordingly, since the universe had developed first dynamically, then chemically and, finally, biologically, 'the abstract sciences, physics and chemistry, must precede and form the basis of the concrete science, mineralogy.'(27)

In a paper which is fundamental to understanding his approach, 'The chemical and mineralogical relations to metamorphic rocks', (28) Hunt claimed categorically that in the earlier periods of the earth's history, 'chemical forces of certain kinds were much more active than at the present day'. In successive periods of time, he suggested, chemical alterations in the nature of the oceans and in the detritus carried down to them from the land, had given rise to deposits - precipitates and elastics (i.e. rocks made from broken pieces of igneous or sedimentary rocks). Such deposits, affected by heat, pressure, and the presence of trapped water, had undergone progressive changes in chemical composition which, following metamorphosis, produced some of the puzzling features of petrography, such as the appearance of metamorphosed strata in unaltered formations, or igneous rocks which were chemically similar to Precambrian and Paleozoic sedimentary rocks. (29) Hunt's audacious suggestion and hope was that this hydrothermal theory would provide a chronometer for the chemical dating of a given series of stratified rocks possessed of igneous intrusions, in the absence of other fossil or stratigraphic evidence. In fact, such a chemical chronometer was doomed to failure, for apart from the introduction of such twentieth-century techniques as rates of chemical sedimentation and radioactive dating, geochemists have been unable to make reliable correlations of the ages of rock specimens and their mineralogical composition. (30)

These preliminary remarks have served to demonstrate that whereas Forbes followed experimental, microscopic, physical and chemical evidence from geological samples to reach petrographical conclusions, Hunt mainly drew his deductions from theoretical chemical models whose ultimate premises lay outside the atomic -molecular tradition. In the manner of the biblical Paracelsians, he believed that the course of geological history could only be determined by the chemist. As the historian of the Canadian Geological Survey has commented: Hunt 'used his chemical skills to solve theoretical problems, to explain how things <u>could</u> have happened (which often transformed in his mind into how they <u>did</u> happen)'. (31) Like James Hutton, Hunt required a history of the earth which formed a sub-discipline of a unified system of nature; hence sympathy for his geological and chemical programmes really required an understanding of, and commitment to, his philosophy of science. (32)

Forbes's criticisms of James Geikie

In February 1867 David Forbes criticised two recent papers by James Geikie who, like his older brother, Archibald, was then working on the Geological Survey of Scotland. (33) According to Forbes, Geikie had shown himself too much the geologist, and insufficiently 'at home either in chemistry, mineralogy, petrology, or physics'. This unfamiliarity accounted for Geikie placing so much emphasis upon 'crude' field observations and 'undervaluing the all-important assistance to be derived from a knowledge of the collateral sciences'. (34) As Forbes explained, there was nothing personal in his attack (he had never met Geikie); rather, he was trying to make a methodological point about the way field geologists paid lip service to chemistry, mineralogy and petrology, while possessing very little systematic knowledge of these subjects. In particular, Forbes found Geikie's use of chemistry 'incorrect', (35) and his command of petrographical evidence dogmatic yet hopelessly unsystematic.

Geikie's reply is relatively unimportant except that he cited Hunt's papers as the origin of his dogmatic petrogenetic speculations. (36) Its effect was to turn Forbes's attention away from the immediate methodological and pedagogical point that he was making, to the petrographical speculations of Hunt. In a further reply to Geikie in May, Forbes argued strongly for a fixed scheme of petrological notation, and while not wishing to reject the possibility of the kind of hydrothermal action favoured by Hunt, he objected that it was an undemonstrated supposition. Fortunately, however:

> Since my former communication, the arrival of Dr. Sterry Hunt in this country has procured me the pleasure of his personal acquaintance. The opportunity thus afforded us, of comparing notes on chemical geology, showed how many similar conclusions we had respectively come to, from the study of widely different parts of the globe, and assured us that any difference in opinion could not arise as to the agencies employed in Nature's operations, although we might be somewhat at variance as to the precise extent to which each agent had been engaged. (37)

Hunt's Royal Institution lecture

Hunt visited England in May 1866 shortly after witnessing some of Henri Sainte-Claire Deville's experiments on chemical dissociation in Paris. (38) On 31 May he lectured on 'The Chemistry of the Primeval Earth' at the Royal Institution, a stenographic report of which appeared in the Chemical News, followed by the author's corrected version in the Geological Magazine and the Proceedings of the Royal Institution. (39)

Hunt's starting point in this astonishing and brilliant lecture, was the nebular theory of the earth's origin, together with the new science of spectroscopy and Deville's work on gaseous dissociation at high temperatures. Considering the earth as a gradually cooling gaseous mass separated from a nebulous body, Hunt went on to describe a primary differentiation of the material into a series of chemical compounds, each stable at successively lower temperatures. These condensed as liquids at the centre (arranged according to their specific gravities) surrounded by the remaining gas. A further reduction of temperature was then postulated as lending to solidification from the centre in accord with the model of the earth proposed for dynamic reasons by mathematicians and physicists such as William Hopkins and William Thomson. The solid centre took no further part in the production of superficial rocks, which had formed from an outer skin and a liquid bath of no great depth below. The original primitive crust being now buried 'beneath its own ruins', Hunt attempted by chemical reasoning to reconstruct it.

If the present solid land, together with the sea and atmosphere were to react upon one another under the influence of intense heat, it would result in 'the conversion of all carbonates, chlorides and sulphates into silicates, and the separation of the carbon, chlorine and sulphur in the form of acid gases', which in the presence of an atmosphere of nitrogen, water vapour, and a probable excess of oxygen, would produce a dense 'primeval atmosphere'. The resultant fused mass, which would contain all the bases as silicates would, Hunt suggested, have resembled in composition certain furnace slags and volcanic glasses. Moreover, since the atmosphere would have been so very dense, under the pressure of such a high barometric column, condensation could have occurred at a temperature much above the present boiling point of water. In that case, 'the depressed portions of the half-cooled crust would be flooded with a hot solution of hydrochloric acid whose action in decomposing silicates [was] easily intelligible to the chemist'. Such a process resulted in the formation of chlorides and the separation of silica (as quartz) and sea-water containing sodium chloride and magnesium chloride, together with salts of aluminium and other bases. The atmosphere, now deprived of chlorine and sulphur, would thus differ from the modern atmosphere only in its higher content of carbonic acid.

The second phase of the reaction of the atmosphere with the primitive crust, Hunt speculated, had been subaerial - the formation of clays and soluble carbonates which would have precipitated the dissolved alumina and heavy metals in the sea, and the decomposition of calcium chloride into calcium and sodium chloride. In support, Hunt quoted the composition of sea waters trapped in the pores of older stratified rocks which were high in salts of calcium and magnesium. Hunt connected the removal of carbon dioxide from the primeval atmosphere with changes in organic life by citing both the chemical precipitation of limestones and the effect of plant respiration. He also alluded to Tyndall's experiments on the effect of carbon dioxide in preventing radiation of heat, for this might explain the occurrence of tropical climates in what are now temperate and arctic zones. Hunt also argued that the carbonic atmosphere would have favoured the formation of gypsum and magnesium limestones, which he claimed to have verified by experiment.

In the final section of his lecture, Hunt considered the effect of heat from below upon the buried sediments, 'converting them into what are known as crystalline or metamorphic rocks, such as gneiss, greenstone, granite, etc.', going on to suggest that granite was always derived from altered sediments because it included in its composition quartz - which he supposed could 'only be generated by aqueous agencies . . . at comparatively low temperatures'. Gases were disgorged when sediments were heated in the presence of water, giving a rational chemical explanation of volcanoes. Hunt thus considered 'that all volcanic and plutonic phenomena have their seat in the deeply buried and softened zone of sedimentary deposits of the earth and not in its primitive nucleus'. (40) In a final flourish, he alluded to the old controversy between plutonists and neptunists, and suggested, reasonably enough, that his model rendered justice to both sides.

> We have seen how actions dependent on water and acid solutions have operated on the primitive plutonic mass, and how the resulting aqueous sediments, when deeply buried, come again within the domain of fire, to be transformed into crystalline and so-called plutonic or volcanic rocks.

Forbes's reaction

As we have mentioned, Hunt's stimulating 'crenitic hypothesis' -as he later called it - was widely publicised. In October 1867

David Forbes contributed rejoinders to both the Geological Magazine and to the Chemical News - explaining in the latter that he had received a personal invitation from Hunt to publish his objections to Hunt's interpretation in the form of 'a friendly fight'. (41) In this way Forbes attempted to bring his criticisms forward to both the chemical and geological communities. The reply in Chemical News concentrated on the chemical aspects of Hunt's lecture, that in the Geological Magazine was both geological and more general.

The first point in Hunt's model attacked by Forbes on empirical grounds was the theory of the cooling of the earth in which the globe solidifies from the centre, being now solid to the core. With his superior knowledge of metallurgy, he was able to cite the production of castings in factories and the manufacture of bullets as indicating the contrary. He also questioned the assumption of a constant indefinite increase of fusing points by pressure, suggesting that in reality (as Hopkins had already shown) there would be a limiting value for every substance. Like many mid-Victorian geologists, Forbes was unhappy with the solid core model of the earth, and he felt there was sufficient observational evidence to suggest that 'the earth does enclose a vast reservoir or reservoirs of still fluid igneous matter in its interior'. (42)

In his views on the chemical composition of the atmosphere of the cooling globe, Forbes again differed strikingly from Hunt. Sulphur combined with heavy metals to form dense sulphides rather than to form an atmosphere of sulphurous acid; chlorine combined chiefly with the alkali metals to form salts rather than to create an atmospheric hydrochloric acid from the 'mutual reactions of sea-salt, silica and water'; (43) equally unsatisfactory was Hunt's explanation of the saltiness of the sea by the action of 'a highly-heated acid deluge' which removed sodium from sodium silicate, leaving quartz behind.

What then were Forbes's own views on the chemistry of the primeval earth? In Forbes's model the earth had begun as a molten sphere surrounded by a gasiform atmosphere, both of which 'were composed of concentric layers or zones of different densities and chemical composition'. (44) The molten solidising zones were a dense sphere of 'metallic bodies more or less combined with sulphur, arsenic, etc.', surrounded by a basic silicate layer and an external crust of acid silicates. The atmospheric zones comprised vaporised salts, carbon dioxide, steam, oxygen and nitrogen. Condensation of this atmosphere produced the salt ocean.

> From this stage in the earth's history, the author [Forbes] believes that all the changes which have taken place in the globe, up to the present time, have been effected by agencies similar to those going on in it at this present day; rocks were formed from the wearing down and disintegrating action of the atmosphere upon

> the primitive crust, and the subsequent stratification of the debris, so formed by the action of the sea; just as they are at present in the course of formation from the disintegration of pre-existing rocks. Eruptions of igneous matter from the still fluid interior from time to time disturbed and broke through the primitive crust and the rock strata above it, in course of formation from its debris, just as at the present day (though possibly on a somewhat smaller scale), similar outbursts are produced by volcanic action. The products of such older eruptions are almost identical, in chemical composition, with those of the newer period. (45)

Moreover, Sorby's microscopic examination of thin sections of limestone had proved them to be the 'debris of organisms' and not originating in chemical precipitation. Hunt's neglect of Sorby's research (as lamentable a lapse in Forbes's eyes as Geikie's ignorance of chemistry) had also led him to impossible mechanisms for the origin of dolomites and magnesium limestones.

Forbes's fundamental geological difference with Hunt was over the question of the origin of granite, which he viewed as degraded igneous material which had been reconsolidated by pressure. By again drawing on his friend Sorby's empirical investigations, by defining similarities between acid volcanic lavas and granites, Forbes attacked Hunt's claim that quartz could be formed only be sedimentary processes from the quartz-free primitive crust, and his belief that granites were derived solely from the reworking of non-igneous sediments. The microscope, not chemical analysis, was the arbiter of whether a rock was igneous or metamorphic.

Hunt's reaction

Hunt was no stranger to scientific debate. Always willing to defend his views vigorously, he replied from Canada in both the *Chemical News* and the *Geological Magazine*. But we shall not examine Hunt's replies in detail here. It is sufficient to notice that Hunt reiterated his model and found objections to Forbes's alternative. In particular, he stood by the sedimentary origin of granite, and challenged Forbes to explain 'the intervention of water in all igneous rocks which, as he declares, are outbursts from the still fluid interior of our globe'. (46) For what follows, it is worth noting especially Hunt's claim for his Royal Institution lecture as a synthesis 'of modern investigations in physics, chemistry, mathematics and astronomy' from which he had constructed 'a scheme which would explain the development of our globe from a supposed intensely heated vaporous condition down to the present order of things'. (47)

Forbes on Hunt's geological chemistry

On 20 January 1868 Forbes composed a second, more elaborate critique of what he now labelled Hunt's 'geological chemistry' for Chemical News which, on his insistence, also appeared in the Geological Magazine. (48) The details of Forbes's nine-point rejection of Hunt's model need not detain us, though it is worth emphasising that Forbes continually drew upon experimental (especially metallurgical) evidence to refute, or to cast doubt upon, Hunt's speculations.

The protagonists' care in appealing to both the chemical and geological communities aroused great interest. At the end of his second reply, Forbes was able to announce that he had been invited by the Council of the Chemical Society to lecture on the substance of the debate. Forbes's lecture to the Chemical Society, 'On Chemical Geology', (49) on 20 February 1868, will inevitably remind the historian of Brodie's lecture on the calculus of chemical operations of the year before, and of Williamson's reply in 1869 (50) - especially since Williamson was also in the Chair on this occasion. Indeed, the audience, and the degree of interest generated were strikingly similar:

> There was an unusually full attendance of Fellows, and several guests, amongst whom were Sir Roderick Murchison, Professor John Morris, and other distinguished members of the Geological Society. The limited accommodation available in the meeting room [in Old Burlington House] was altogether insufficient to provide for the large audience that attended. (51)

Of course, the other principal involved, Sterry Hunt, was not present, and to that extent the lecture lacked the polarising drama of the atomic debates. Nevertheless, the discussion of Forbes's paper - which largely reiterated his own model of the earth's chemical history and urged a fusion of what he described, like Hunt, of neptunist and plutonist views - was animated. Neither water nor fire were sufficient modes of explanation of geological history; heat, electricity, light and pressure must all be considered within the theoretical structure. Most of the discussants seem to have agreed with Forbes, though Murchison (significantly an older field geologist) felt incompetent to judge, and only the aged William McDonald, Professor of Natural History at the University of St Andrews, who appeared to think that Hunt was reviving Werner's system, supported Hunt whole-heartedly.

It was left to a pharmacist, B. H. Paul - a pupil of Liebig's and the translator of C. G. C. Bischof's enormous Chemical and physical geology (52) - to ask rhetorically at the meeting what were the broad principles upon which chemists should investigate geology. Forbes's answer came in a third critical reflection on Hunt, which again appeared in two slightly different versions for chemists and geologists. (53)

In considering the relationship between geology and chemistry, Forbes concluded, the inquirer must decide which science formed 'the basis, or starting point', since this could 'not fail to exercise an important influence on his reasonings and deductions'. (54) The fundamental reason for their disagreement, suggested Forbes, was their different starting points - his own of chemical geology, and Hunt's of geological chemistry. Whereas Forbes began as a geologist and then applied 'chemistry, especially experimental chemistry, to the explanation of known geological phenomena', Hunt started 'from data purely chemical, and then looks around for geological instances to which they may be applied'. For example, knowing that sodium carbonate reacts with calcium chloride to precipitate calcium carbonate

$$Na_2CO_3 + CaCl_2 \longrightarrow CaCO_3 + 2NaCl$$

Hunt asserted that all calcareous earths were formed by similar precipitations. Hunt's approach, while ultimately rooted in experimental science was, in practice, Forbes angrily concluded, a juxtaposition of selected quotations from old authorities and disjointed chemical fragments, whereas the only valid approach was from experimental geology.

The tone of the debate

This was the last paper which contributed directly to the controversy, for although Forbes contributed a fourth essay to the July number of the Popular Science Review, which was substantially reprinted in the Chemical News on 23 October 1868, (55) it did not refer directly to Hunt's papers. Unfortunately Hunt, who had returned to Canada, never took up Forbes's further challenges, and ignoring his position he continued as before to publish and to republish papers on volcanic action, igneous and metamorphic petrology, and even on celestial chemistry. (56)

The abrupt end of this exchange of views can in part be ascribed to the increasing element of personal abuse which was injected into the discussion. Hunt's original invitation to Forbes had been for a 'friendly fight'. Forbes had readily agreed,

> in the belief that fair discussion advances science, by developing energetically both sides of the question; such discussions should, however, only be indulged in by those who can give and take with equal good grace, without losing temper or deviating from the subject at issue by indulging in recriminations or personalities . . . which unfortunately sometimes creep into discussion even in this enlightened age. (57)

Since Hunt's temper was only too easily aroused, however, Forbes's exploitation of the affair to awaken 'the study of chemical geology from its present semi-torpid state in England',(58) meant that the debate, despite its one-sided character, immediately came to involve 'recriminations', 'personalities', and to assume an ominous belligerent tone.

There were two obvious presuppositions behind Forbes's raillery and insults. First, that despite Hunt's expertise and experience of the geology of the vast Canadian continent, he had no mandate to generalise his experience to include European geology - and from there, to cosmic geology.

> Europe differs greatly from Canada, and, amongst other things, in close competition being the order of the day . . . If he speaks as a geologist, it may fairly be inquired whether he considers his Canadian experience sufficient to enable him to arrive at such sweeping generalisations. Sir Charles Lyell has stated that three things were essential to a geologist, namely, 'to travel, to travel, and to travel'; and such advice may be recommended to Dr Sterry Hunt before he ventures again to generalise for the world on the strength of a local knowledge of a minute part of the same. (59)

Forbes was, of course, more widely travelled - it appears he had even visited Canada briefly (60) - though as Hunt retorted, 'those who make many pilgrimages rarely become saints'. (61)

This retort was, in fact, a quotation from Thomas à Kempis - which brings us to Forbes's second presupposition. From the tone and kind of literary allusions made by Hunt, and above all, by his appeals to literary authorities, Forbes clearly came to suspect Hunt of Roman Catholicism. (62) In June 1868, Pius IX had convened an Oecumenical Council in Rome which was to assert the doctrine of Papal Infallibility and to define the relations of Roman Catholicism to modern science in terms as uncompromising as the *Syllabus errorum* of 1864. It is therefore perhaps not surprising that Forbes adopted a 'warfare' image of his dispute with Hunt or that he should claim that Hunt's 'authoritarianism' would restrict the progress of geology.

> It is well known that a knowledge of the world acquired by travel is the best antidote to bigotry or one-sided opinions. What we require are geologists not saints, and although it may be that in Canada geologists are esteemed in proportion to their saintly pretentions, experience on this side [of] the Atlantic does not tend to prove that any of the natural sciences have been as yet much advanced by the labours of the would-be-saintly portion of the community . . . [Hunt's] idea of scientific warfare consists in an attempt to overwhelm and crush his opponent with sneers and countless accusations of incompetency and ignorance; ignorance of chemistry, of geology, of petrology, mineralogy, microscopy, literature of the subject, etc., whilst at the same time he does not fail to herald in his own views as what might be termed the quintessence of the combined 'results of modern investigations in physics, chemistry, mathematics and astronomy'. (63)

Discussion

Hunt's openly confessed debt to Abraham Werner was bound to make his geological theories suspect in the eyes of British geologists who had been brought up on Lyell, (64) though his emphasis on the role of water in the formation of eruptive rocks is a tradition which has remained in Canadian geology and proved to be of theoretical importance.

In part the differences between Hunt and Forbes over the mode of origin of granite can be put down to the difference of the areas with which they were most familiar, Hunt having developed his ideas in the context of an ancient shield-area, whereas Forbes's experience was derived especially from the consideration of areas of more recent orogeny, but the contrast between Hunt's chemical and mineralogical background with that of Forbes in metallurgy and smelting practice, together with Forbes's strong bent for Sorby's microscopic examination, was clearly of great importance in the development of their respective positions. The debate seems a good example of what R.G.A.Dolby has called 'geographical variation in conceptions of the problem field of science'. (65) Despite several common allegiances, their upbringing in different continents, their different formative influences, and their different scientific and philosophical experiences, reading and travelling styles, led them to divergent approaches to geology.

Forbes analysed their differences clearly when he remarked that whereas he looked to chemistry only to find an explanation or insight into a geological problem, Hunt began with a chemical reaction and from it deduced the necessary geological situation. Despite an active surveying career with Logan that was far from usual for a Survey chemist, Hunt was not primarily a field geologist. The old adage that 'the best geologist is the one that has seen the most rocks' is, of course, particularly applicable to those who propose very general hypotheses.

Forbes's position was remarkably similar to that of the Swedish chemist, Berzelius, who (as Tore Frängsmyr has pointed out):

> objected that chemistry cannot give an explanation of all geological observations. If the chemist gives his views of the building of the Alps, he must first ascertain from the geologist how the process of building has worked. If the geologist can tell this story in a satisfactory way, then the chemist may be able to explain the chemical side of the process. (66)

Like the Greek astronomer who had first to consult the philosopher-physicist for the principles upon which the phenomena could be saved, the chemical geologist had to begin with conjectures proper to geology before applying his specialised chemical knowledge; otherwise, like Aristarchus in astronomy, he would invent chemical systems which bore little relation to the real historical or present geological situation. Indeed, Forbes felt that 'no chemico-geological investigations or chemical analyses of British rocks or of their component minerals have as yet been made which could serve as a basis for . . . generalisation about "the nature and formation of our metamorphic and eruptive rocks" '.(67)

This assertion of the primacy of geology over chemistry was in interesting contrast to the secondary role which British geologists were content to assume with respect to the physicists whom they allowed to dictate the chronology of the earth. (68) Although Forbes did not allude to Kelvin's assumed omniscience in geology, it is tempting to conclude either that geologists felt that physicists possessed more authority than chemists, or that Forbes (if not other geologists) was determined that geologists would not be dictated to by chemists as they had been by physicists. Of course, the gulf between Hunt and Forbes really went much deeper, for Hunt's conception of chemical processes was totally different from the atomic-molecular tradition accepted by Forbes. Whether the latter realised this is, however, rather doubtful.

In conclusion, two other matters deserve mention, for the debate had implications for the education of geologists and for the future emergence of the discipline of geochemistry.

In a letter to Faraday commenting on the York meeting of the British Association in 1844, Liebig expressed amazement at the way geologists hogged the proceedings despite their scientific ignorance.

In England, it seemed to a German chemist, one could be a 'great' geologist 'without [having] a thorough knowledge of Physics and Chemistry' or mineralogy. (69) Even allowing for exaggeration and for Liebig's pro-chemist viewpoint, his comment is perceptive, and is confirmed by other sources. The 'collateral sciences' impinged on geology in two ways - through palaeontology, where a knowledge of botany and zoology was necessary; and through petrology and mineralogy, where a knowledge of chemistry and physics was needed. Now, although mineralogy had received much attention from the pupils of Werner and the founding-fathers of the Geological Society, with 'the decline of Wernerianism the study of minerals gradually lost its hold . . . and younger men . . . cared less for the inorganic than for the organic treasures of the rocks'. (70) As the questions for the Cambridge Natural Science Tripos (from 1851) suggest, the taste for mineralogy was replaced by palaeontology. (71) The knowledge of natural history informally gained from field and sporting pursuits by country gentlemen or public schoolboys was for several decades a sufficient introduction to biology for the British geologist faced by palaeontological mysteries. If, and when, chemical information about rocks and soils was required, then (as we have previously suggested) the analytical chemist could be consulted. Consequently, the initial thrust of what we might term geochemistry came from the chemists rather than the geologists - a situation which Forbes clearly deplored. Ironically, just as (through the efforts of Forbes, Sorby and others) geologists were amputating geology from the amateur tradition of natural history and forcing it into the family of professional, physical, laboratory sciences, most chemists (despite Hunt) were tending to ignore geology for what appeared to be more exciting liaisons.

Although the word 'geochemistry' was used as early as 1838 by the chemist Schonbein, the term 'chemical geology' was much older. Until the twentieth-century development of a geochemistry based upon the application to geology of chemical techniques such as X-ray analysis, optical spectrography and isotope studies, the two terms were usually used synonymously.

But if the distinctions now drawn between 'geochemistry' and 'chemical geology' (as, for example, in the two modern journals with these respective titles) are similar to those behind Hunt's geological chemistry and Forbes's chemical geology, (72) why did Hunt's geological chemistry fail to lead immediately to a research programme in geochemistry? The importance of later techniques and chemical concepts have already been mentioned, but a more significant factor was that Hunt's system was a 'cosmic chemistry', a speculative system whose falsifiability lay beyond the analytical resources of his time. Indeed, what did interest some chemists in his scheme was not its implications for geology, but its suggestion of evolving chemical elements and compound atoms which had been given independent credence by contemporary organic chemistry and the periodic law.

Because Liebig's system of microcosmic animal chemistry (physiological chemistry), which was as speculative as Hunt's evolutionary and crenitic hypotheses, was immediately experimentally falsifiable, it stimulated the systematic work which Hoppe-Seyler institutionalised as the discipline of biochemistry in the 1870s. However, even if Hunt's particular geochemistry had been falsifiable in principle, it would have received short shrift from the majority of the chemical community because this community rejected his starting assumption that the atomic theory was totally false.

Sociologists of science have examined many cases of contemporary migrations from one discipline to another and the emergence of new cross-disciplines like radio-astronomy, (73) but their case studies have all been success stories. What are the necessary and sufficient conditions for interdisciplinary success? Although in the case of geochemistry further research will be necessary, the discipline does tend to support the findings of Ben-David and Edge and Mulkay that activity in an applied field adjacent to research in the pure sciences is significant. By the 1860s, few chemists, apart from Hunt, were interested in geology *per se*. In chemistry itself the breakthrough in structural organic chemistry was providing a rich harvest of interesting problems, while periodicity and the new physical chemistry offered other tempting avenues of research. For the more interdisciplinary chemist of the 1860s and 1870s there were interesting problems in photography, spectroscopy and biochemistry. Institutionally, geologists were divorced from chemists in the fast-developing universities and university colleges of the 1870s, and the only meeting grounds for chemical-geological dialogue remained, as with Hunt, in the applied setting of the Geological Surveys and their associated museums. It was here, and not in the chemistry laboratory or the university, that in the early 1900s Hunt's successors, F. D. Adams (Canadian Geological Survey), F. W. Clarke (U.S. Geological Survey) and the Swiss-Norwegian mineralogist, V. M. Goldschmidt began to forge the craft and science of geochemistry. By then chemistry had become as historical as geology - as Hunt had speculated; and geology had become as mindful of the physical sciences as it had once of natural history - as Forbes had demanded. (74) Armed with the periodic law, ideas of nucleosynthesis and with X-ray machines, an experimentally-based chemistry of the earth was possible and neither the geological chemist nor the chemical geologist would dispute the other's emphasis.

Chronological bibliography of the Forbes-Hunt papers

Geikie, J. 1866a. 'On the metamorphic lower Silurian rocks of Carrick, Ayrshire', Quart. J. Geol. Soc. Lond., xxii, 513-34.

Geikie, J. 1866b. 'On the metamorphic origin of certain granitoid rocks and granites in the Southern Uplands of Scotland', Geol. Mag., iii, 529-34.

Forbes, D. 1867a. 'On the alleged hydrothermal origin of certain granites and metamorphic rocks', Geol. Mag., iv, 49-59.

Geikie, J. 1867a. [First reply to Forbes], Geol. Mag., iv, 176-82.

Forbes, D. 1867b. [First reply to Geikie], Geol. Mag., iv, 225-30.

Geikie, J. 1867b. [Second reply to Forbes], Geol. Mag., iv, 287-8.

Hunt, T. S. 1867. 'On the chemistry of the primeval earth', Chem. News, xv, 315-17; (reprinted in D. M. Knight [ed.], "Classical scientific papers. Chemistry. Second series", London, 1970, pp. 364-6). The essay also appeared in Geol. Mag., iv, 357-69, with errata 432, 478, 525-6; Proc. Roy. Inst., 1869, v, 178-85; and in Hunt, "Chemical and scientific essays", Boston & London, 1875, pp.35-48. Different shorthand writers were employed by Chemical News and Geological Magazine, and there are material differences in the two accounts.

Forbes, D. 1867c. 'On some points in chemical geology', Chem. News, xvi, 175-7.

Forbes, D. 1867d. 'On the chemistry of the primeval earth', Geol. Mag., iv, 433-44.

Hunt, T. S. 1868a. 'On the chemical geology of Mr. David Forbes', Chem. News. xvii, 27-9.

Hunt, T. S. 1868b. 'A notice of the chemical geology of Mr. D. Forbes', Geol. Mag., v, 49-59.

Forbes, D. 1868a. 'On some points in chemical geology. No. II. Dr. Sterry Hunt's geological chemistry', Chem. News, xvii, 39-41.

Forbes, D. 1868b. 'On some points in chemical geology', Geol. Mag., v, 92-8.

Forbes, D. 1868c. 'On chemical geology', J. Chem. Soc., xxi, 214-41.

Forbes, D. 1868d. [Abstract and discussion of 1868c], Chem. News, xvii, 105-7.

Forbes, D. 1868e. 'On some points in chemical geology. No. III. Dr. Sterry Hunt's geological chemistry', Chem. News, xvii, 111-13.

Forbes, D. 1868f. 'On Dr. Sterry Hunt's geological chemistry', Geol. Mag., v, 105-11.

Forbes, D. 1868g. 'On some points in chemical geology, IV', Chem. News, xviii, 191-4.

Forbes, D. 1868h. 'On some points in chemical geology', Geol. Mag., v, 366-70; (originally published in Pop. Sci. Mon., July 1868).

Notes

1 A. J. Meadows, "Early solar physics", Oxford, 1970; N. G. Coley, "From animal chemistry to biochemistry", Amersham, 1973. Sociologists of science have shown little interest in the historical phenomenon of interdisciplinarity. See, however, M. J. Mulkay, 'Conceptual displacement and migration in science: a prefatory paper', Sci. Stud., 1974, iv, 205-14; idem, 'Three models of scientific development', Socio. Rev., 1975, xxiii, 509 -23; A. J. Meadows, 'Diffusion of information across the sciences', Interdisciplinary Sci., 1976, i, 259-67; D. O. Edge and M. J. Mulkay, "Astronomy transformed. The emergence of radio astronomy in Britain", New York, London, Sydney & Toronto, 1976. Relevant, but not available to me, is G. Lemaine et al. (eds.), "New perspectives on the development of scientific disciplines", Paris, 1977.

2 H. Spencer, 'On the genesis of science', in "Essays on education", Everyman Books edn., London, 1911, p.295.

3 D. E. Allen, 'The lost limb: geology and natural history', this volume. See also D. R. Stoddart, 'That Victorian science: Huxley's physiography and its impact on geography', Trans. Inst. Brit. Geographers, Nov. 1975, 17-40.

4 P. J. Weindling, 'Geological controversy and its historiography: the prehistory of the Geological Society of London', this volume.

5 Forbes 1868h, p.366. References of this form refer to the chronological bibliography of the Forbes-Hunt papers.

6 R. Rappaport, 'Lavoisier's theory of the earth', Brit. J. Hist. Sci., 1972-3, vi, 247-60; D. McKie (ed.), "Notes from Doctor Black's lectures on chemistry 1767/8", Imperial Chemical Industries, Wilmslow, 1966; J. J. Berzelius, "An attempt to establish a . . . system of mineralogy", London, 1814; T. Thomson, "A system of chemistry", 5th edn., London, 1817, vol.iii; R. Siegfried & R. H. Dott Jr., 'Humphry Davy as geologist, 1805-29', Brit. J. Hist. Sci., 1976, ix, 219-27. Note also D. M. Knight, 'Chemistry in palaeontology. The work of James Parkinson (1755-1824)', Ambix, 1974, xxi, 78-85.

7 G. K. Roberts, 'The establishment of the Royal College of Chemistry: an investigation of the social context of early-Victorian chemistry', Hist. Stud. Phys. Sci., 1976, vii, 437-85; see p.451.

8 J. R. Bentley, 'The chemical department of the Royal School of Mines. Its origins and development under A. W. Hofmann', Ambix, 1970, xvii, 153-81; G. P. Merrill, "The first one hundred years of American geology", reprint, New York, 1964; M. Zaslow, "Reading the rocks. The story of the Geological Survey of Canada 1842-1972", Toronto, 1975.

9 J. D. Burchfield, "Lord Kelvin and the age of the earth", New York, 1975, p.96.

10 Cf. the debate between physicists and geologists over the age of

the earth, Burchfield, op.cit. (9).

11 My interest in the exchange between Forbes and Hunt was first aroused in 1964, when it entered the periphery of my research on the debates over atomism in the 1860s. Subsequently, Nigel C. Griffin, then an undergraduate geologist at the University of Leicester, made the Forbes-Hunt dispute the subject of a History of Science sessional essay. I am grateful to him for allowing me to draw on some of his material. I have also benefited from seeing an unpublished essay (1977) on Hunt's chemical philosophy by Thomas E. Fogg of the Institute for the History and Philosophy of Science and Technology, University of Toronto.

12 J. Morris, 'Memoir of the late Mr. David Forbes', J. Iron Steel Inst.,1876, 521; W. J. Harrison in "Dictionary of national biography". Note also V. L. Hilts, 'A guide to Francis Galton's *English Men of Science*', Trans. Amer. Phil. Soc., 1975, lxv, 48-9.

13 F. Greenaway,'Percy', "Dictionary of scientific biography". Percy was FGS in 1851, in which year he became lecturer in metallurgy at the Royal School of Mines in London.

14 D. Forbes, "Correspondence with Lord John Russell and memoranda", London, 1861.

15 D. Forbes, 'On the causes producing foliation in rocks', Quart. J. Geol. Soc. Lond., 1855, xi, 166-85; 'On the occurrence and chemical construction of some minerals from the south of Norway', "Report of the . . . British Association for the Advancement of Science, 1854", London, 1855, pp.67-8; 'On the igneous rocks of South Staffordshire', "Report of the . . . British Association for the Advancement of Science, 1865", London, 1866, pp.53-6.

16 N. Higham, "A very scientific gentleman. The major achievements of Henry Clifton Sorby", Oxford, 1963, p.43; Higham, p.46, shows that Forbes was on one of his frequent visits to England. For Forbes's championship of Sorby, see 'The microscope in geology', Geol. Mag., 1867, iv, 511-19. It was Forbes's examination of 'a new earth like Thoria and Zirconium' in 1868 which led Sorby into one of his few blunders, the 'Jargonium fallacy', (Higham, pp.92-7).

17 The Times, 12 Dec. 1876, col. 6e.

18 D. Forbes, 'Applications of the blowpipe to the quantitative determination or assay of certain minerals', Chem. News, 1867, xv, 38, 165, 177, 281; 'On the nature of the earth's interior', Geol. Mag., 1871, viii, 162-73; Forbes, 1867c, 1867d.

19 J. Douglas, "A memoir of Thomas Sterry Hunt", Philadelphia, 1898; F. D. Adams, 'Biographical memoir of Thomas Sterry Hunt', Biogr. Mem. Nat. Acad. Sci., 1934, xv, 207-38, which draws heavily on Douglas; W. H. Brock, 'T. S. Hunt', "Dictionary of scientific biography". One may concur with Zaslow, op.cit. (8),p.545 that Hunt warrants a definitive biography.

20 Hunt's role is discussed by Zaslow, op.cit. (8), pp.75-7, 102-3.

21 T. S. Hunt, "Petroleum, its geological relations", Montreal, 1865.

22 Both Hunt's "A new basis for chemistry", Boston, 1887, and "Systematic mineralogy based on a natural classification", New York, 1891, were ignored by the scientific community.

23 T. S. Hunt, 'Considerations on the theory of chemical changes', Amer. J. Sci., 1853, xv, 226-34, and Phil. Mag., 1853, v, 526-30, reproduced in "Chemical and geological essays", Boston & London, 1875, pp.426-37. (The latter will be cited as "Essays".) Hunt did not adopt Oken uncritically. See 'On the objects and method of mineralogy', Amer. J. Sci., 1867, xliii, 203-7, reprinted in "Essays", pp.453-8.

24 J. B. Stallo, "General principles of the philosophy of nature", Boston, 1848, quoted in Hunt, 'Thoughts on solution and the chemical process', Amer. J. Sci., 1855, xix, 100-3, reprinted in "Essays", pp.448-52 (450).

25 Thus what Dumas called 'metalepsy', or double decomposition, proceeded via the paradigm $A + B \rightarrow X \rightarrow C + D$, even though X was not isolated.

26 See P. Baxter, 'Natural law versus divine miracle: the Scottish evangelical response to _Vestiges of the natural history of creation_', Brit. J. Hist. Sci., forthcoming.

27 Hunt, 'On the objects . . . of mineralogy', op.cit. (23).

28 T. S. Hunt, Quart. J. Dublin Geol. Soc., 1864, x, 85-95; reprinted in "Essays", pp.18-34.

29 Cf. Zaslow, op.cit. (8), p.77.

30 Hunt's dating method, which is not discussed by Burchfield, was chiefly attacked and dismissed by Dana.

31 Zaslow, op.cit. (8), pp.102-3.

32 R. Grant, 'Hutton's theory of the earth', this volume.

33 See Forbes 1867a on Geikie 1866a, 1866b.

34 The phrase 'collateral sciences' (usually meaning physics and chemistry as adjuncts to another discipline) was widely used in the 1840s and 1850s, and probably emanated from the broad character of the London University medical degree course. See N. G. Coley, 'The collateral sciences in the work of Golding Bird (1814-1854)', Med. Hist., 1969, xii, 363-76. Forbes had used the expression in supporting Sorby. See "Report of the . . . British Association for the Advancement of Science, 1865", London, 1866, p.53.

35 Geikie, for example, had defined magnesia, instead of serpentine, as the cause of the green colouring of rocks, and identified alkalinity as the cause of the hardness of certain rocks, without offering a chemical demonstration of their alkalinity. See Forbes 1867a, p.52.

36 Geikie 1867a, p.180, in which he also hastened to refute Forbes's allegation that his papers represented official survey doctrines. In defence against Forbes's methodological charge, Geikie replied that no one, with the possible exception of Forbes, could possibly

be proficient in chemistry, mineralogy and geology; there would be a natural bias towards one of the Sciences.

37 Forbes 1867b, p.229, note 1. In reply, Geikie 1867b challenged Forbes to write a textbook on petrology. According to The Times, op.cit. (17), Forbes 'left an immense amount of manuscript information which he hoped in later years to make public . . .' I have not attempted to trace this material.

38 W. H. Brock (ed.), "The atomic debates", Leicester, 1967, pp.13, 24-6.

39 Hunt 1867. The lecture was given extempore.

40 Ibid., p.317.

41 'He [Forbes] would . . . have gladly seen the challenge of Dr. Hunt accepted by someone more able than himself; and, consequently, before entering the lists thinks it proper to explain that he has been induced to do so by Dr. Sterry Hunt's special invitation to have "a friendly fight", which he must confess he was not unwilling to accept, especially as in a late discussion that gentleman's opinions, when quoted by one of his admirers [i.e., James Geikie] . . . in opposition to his [Forbes's] views, were put forward with a show of authority which opinions as yet neither generally accepted nor confirmed cannot be entitled to.' Forbes 1867d, p.434.

42 Ibid., pp.436-7. A striking feature of the debate was the protagonists' lack of a sense of relative time between geological epochs.

43 He opposed this on the grounds that the heat needed for such a reaction would cause water to 'evaporate into space'.

44 Forbes 1867d, p.438.

45 Ibid., p.439.

46 Hunt 1868b, p.58.

47 Ibid., p.49.

48 Forbes 1868a and Forbes 1868b. The versions are almost identical.

49 Forbes 1868c.

50 Brock, op.cit. (38).

51 Forbes 1868d, p.105.

52 C. G. C. Bischof, "Elements of chemical and physical geology", 3 vols., Cavendish Society, London, 1854-9.

53 Forbes 1868e, Forbes 1868f. The Geological Magazine version appears to have been toned down by the editor.

54 Forbes 1868e, p.111.

55 Forbes 1868g, Forbes 1868h.

56 T. S. Hunt, 'On the probable seat of volcanic action', Geol. Mag., 1869, vi, 245-51; 'The liquification of rocks', ibid., 1870, vii, 60-1; 'Celestial chemistry', Proc. Cambridge Phil. Soc., 1881, iv, 129-39, reprinted in D. M. Knight (ed.), "Classical scientific papers. Chemistry. Second series", London, 1970, pp.367-9.

57 Forbes 1867d, p.433.

58 Forbes 1867c, p.175.

59 Forbes 1868a, pp.39, 41; Hunt 1868b, p.59.

60 Forbes 1868e, p.112.

61 Hunt 1868b, p.59.

62 I say 'suspect' because it is not clear whether Hunt's conversion to Roman Catholicism was widely known. In fact, he seems to have abandoned formal religion during the 1860s. Forbes himself admitted elsewhere 'much religious bias of thought; but no respect for revealed religion as a base for such a bias'. See Hilts, op.cit. (12), p.48.

63 Forbes 1868e, p.113.

64 A. M. Ospovat, 'The distortion of Werner in Lyell's Principles', Brit. J. Hist. Sci., 1976, ix, 190-8.

65 R. G. A. Dolby, 'Debates over the theory of solution; a study of dissent in physical chemistry in the English-speaking world in the late 19th and 20th centuries', Hist. Stud. Phys. Sci., 1976, vii, 297-404, (esp. p.393).

66 T. Frängsmyr, 'The geological ideas of J. J. Berzelius', Brit. J. Hist. Sci., 1976, ix, 228-36, (229).

67 Forbes 1868h, p.367.

68 Burchfield, op.cit. (9).

69 L. P. Williams (ed.), "The selected correspondence of Michael Faraday", Cambridge, 1971, i, 430.

70 F. W. Rudler, 'Fifty years' progress in British geology', Proc. Geol. Assoc., 1887-8,x, 234-72 (see 257-9). Rudler estimated the decline in petrographical and mineralogical papers published by the Geological Society to be from 42% (1811-21) to 9% (1845-9). For a discussion of the growing 'discipline-divide' between geology and mineralogy during the early nineteenth century, see R. Porter, "The making of geology. Earth science in Britain 1660-1815", Cambridge, 1977, pp.171-6. Note also D. Forbes, 'Researches in British mineralogy', Phil. Mag., 1867, xxxiv, 329-54, esp, p.331 where he blames decline of interest in mineralogy on palaeontology and organic chemistry.

71 Generally speaking, British geologists (unlike their continental colleagues) were not trained in the physical sciences. After 1851 there were a few candidates for the Natural Science Tripos at the University of Cambridge who were biased towards geology and who had previously sat the Mathematical Tripos (which demanded some physics). Most Natural Science Tripos candidates seem, however, to have ignored mineralogical questions. (My thanks to Roy Porter for this information.) Note the foundation of the Mineralogical Society was as late as 1876.

72 According to K. Rankama & T. G. Sahama, "Geochemistry", Chicago, 1950, the chemical geologist has only geological applications in mind; the geochemist is only partly interested in geology as such, being more concerned with the abundance and distribution of elements (ie, with chemistry). See A. A. Manten, 'Historical foundations of chemical geology and geochemistry', Chemical Geology, 1966, i, 5-31, (6-7). Of course, the very issue which interested Hunt philosophically - primeval chemistry - currently

interests chemical geologists as much as geochemists, let alone astrophysicists and biologists.

73 See (1). Historically, there have been failures to create viable inter-disciplines, e.g. J. J. Sylvester's algebraic chemistry (graph theory).

74 H. B. Woodward, 'Notes on modern chemistry and physics', Geol. Mag., 1868, v, 395-9, was written specifically to counter Forbes's accusation that geologists neglected the physical sciences.

The controversy of the Moulin-Quignon jaw: the role of Hugh Falconer

PATRICK J. BOYLAN

The Piltdown 'early man' fraud has been extensively discussed and debated over the past twenty-five years in a range of both scientific and more popular studies, but there has been little modern discussion of the alleged finds at Moulin-Quignon and Menchecourt near Abbeville, northern France, in terrace gravels of the Somme, in 1863, even though 'l'affaire Moulin-Quignon' was in many ways a direct parallel and predecessor of the Piltdown forgeries, as has been pointed out by Kenneth Oakley, who first exposed Piltdown, and Ronald Millar, who examined the similarity between the Moulin-Quignon and Piltdown scandals in his examination of the Piltdown fraud. (1)

The 'planting' of evidence was not of course a new problem of the mid-nineteenth century. Antiquarians in particular had been troubled by frauds, forgeries and dubious evidence for centuries. 'Pious' or 'miraculous' religious frauds were common in most parts of the world, and fraud for financial gain was by no means uncommon in the days when much geological collecting relied mainly on what was sometimes termed 'the silver hammer' - sixpences and shillings (or their local equivalent) offered to impoverished quarrymen for choice cabinet specimens, and the emergence of an active international market in antiquities and in geological and natural history specimens in the second half of the eighteenth century increased the risk of such frauds. In considering alleged frauds in the field of scientific discovery a further possibility is that 'evidence' might be fabricated in order to enhance the reputation of the discoverer or to support a particular argument or viewpoint in a controversy. An even more sinister counterpart of dishonesty of this kind is the deliberate fabrication of evidence in order to discredit an opponent or his work or opinions. Finally, in analysing a scientific fraud or forgery the possibility that the incident began as a straightforward practical joke (which perhaps misfired) must also be borne in mind. As with the Piltdown forgeries, an investigation of the Moulin-Quignon case is far more than a detective story. In both cases the two most likely explanations are the same: deliberate fraud by the inner circle of discoverers or an attempt to discredit the cognoscenti for being convinced, even briefly, by what should have been in each case an easily detectable fraud. However, both offer a less likely third possibility - financial gain by the quarrymen in the case of Moulin-Quignon; a practical joke that went wrong in the case of Piltdown.

The reaction to the discoveries at Moulin-Quignon also throws light on the different national styles of scientific organisation and investigation and, possibly, the effect of political factors in the relationship between Britain and France. The French scientific establishment, represented by the *Institut* and the *Académie des Sciences* had ignored the growing body of evidence on the antiquity of man that the amateurs of Abbeville had been putting forward from the 1830s, in the same way that the similar evidence from the various local workers in Kent's Cavern, Torquay, was not accepted by the English scientific establishment of the Royal Society, the Geological Society, and the British Association between 1826 and 1858. The French scientific establishment was heavily institutionalised compared with the more individualistic tradition of British science, and having only recently accepted the (genuine) evidence of the great antiquity of man in the Somme gravels reacted defensively and bureaucratically to the attack on the authenticity of the Moulin-Quignon 'finds'. The quasi-legal method of investigation initiated by the French institutions presented the Royal Society, which was asked to participate, with a problem in finding a suitable team of British participants. Moreover, in response to the Anglo-French 'Cobden' commercial treaty of 1860, by the time of the Moulin-Quignon affair a conscious effort was being made to improve cultural relations between the two countries. The initial denunciation of the alleged finds by British scientists was clearly regarded as contrary to this new spirit of friendship and co-operation, and the (short-lived) acceptance of the finds as authentic by the British participants in the Anglo-French investigation was regarded as something more than a purely scientific achievement. Another interesting aspect of the affair was the apparently inconsistent attitude of the British workers who attacked the authenticity of the Moulin-Quignon finds, since they were some of the leading advocates of the case for the great antiquity of man, and had argued only four years earlier for the acceptance of the original Somme Valley evidence. There was in fact no inconsistency here: it was precisely because of their convictions in respect of the antiquity of man that they wished to avoid any possibility of association with a forgery which, on exposure, might have set back the advancement of knowledge in this field by a matter of years, possibly decades.

The recent rediscovery and cataloguing of a substantial collection of letters and other papers of one of the principal British geologists involved, Hugh Falconer (1808-1865), in the Falconer Museum, Forres, Moray, Scotland (2) throws new light on the role and attitudes of Falconer in the spring and summer of 1863 as the Moulin-Quignon crisis evolved. Falconer had qualified as a medical doctor in Edinburgh, where he also attended the natural history and geology courses of Jameson, the botany course of Graham and the divinity course of Chalmers. After entering the medical service of the East India Company in 1830, he undertook only a single tour of military duty in India before being appointed to take charge of the Botanic Garden at Suharnpoor in northern India, where his achievements included the successful establishment of tea growing in India. He also quickly established himself as an outstanding explorer and taxonomic botanist, and as one of the world's leading vertebrate palaeontologists. His work on the fossil vertebrates of the Siwalik Hills earned him a world-wide reputation by the time he was twenty-nine years old, with the award of the Geological Society's highest honour, the Wollaston Medal. Returning to England on sick leave in 1842 Falconer carried out a wide range of scientific work, which further enhanced his reputation within the British scientific community, before returning to India in 1847 as Superintendent of the Calcutta Botanic Garden. On his retirement in 1855 he settled in London and continued his scientific work, particularly on the taxonomy of fossil mammals, and played a leading part in the activities of the scientific institutions, notably as a Vice-President of the Royal Society and as Foreign Secretary of the Geological Society.

The collection at Forres includes eighteen original letters written in 1863 from Falconer to his favourite niece, Mrs Grace McCall (a widow who later married Joseph Prestwich (1812-1896), one of Falconer's closest geological collaborators). It seems likely that the whole collection of Falconer Papers at Forres was gathered together by Grace (Lady Prestwich) who seems to have been contemplating a full biography of Falconer (3) when Prestwich died, after which she concentrated her efforts on the biography of her late husband. (4) Grace made some of her own letters available to Charles Murchison for his short biographical sketch, (5) and the collection also contains copies of some of Falconer's most important letters to Joseph Prestwich and John Evans, the archaeologist and geologist. A few quotations from these two series of letters have also been published <u>in passim</u> by Lady Prestwich, (6) and by Joan Evans (7) who found the originals in her father's papers on the death of her step-brother, Sir Arthur Evans.

Altogether in twenty-five of the original and copy letters at Forres, Falconer writes of the Moulin-Quignon controversy: perhaps most revealing are a series of letters to Grace McCall written day by day during the meeting of the Anglo-French commission established by the _Institut_ and the _Académie des Sciences_ to investigate the Moulin-Quignon finds. In addition, Falconer's own writings during the evolving crisis, (8) the official accounts of the Anglo-French commission (9) and a paper on Moulin-Quignon written in the summer of 1863 but not published in his lifetime, (10) taken with the letters at Forres, make possible a reconstruction of Falconer's growing concern and alarm about an event that clearly caused him great personal distress and which could have destroyed public credibility in the abundant genuine finds of very early flint implements in the Somme gravels.

The discoverer of the controversial jaw was Jacques Boucher de Crevecoeur de Perthes, who was born in Rethel, France, in 1788. He moved to Abbeville as a child, and it was there that he eventually succeeded his father as the local _Directeur des Douanes_ in 1825. From that year he remained in Abbeville, dividing his time between public administration and private study and writing until his death in 1868. Although best known for his important work as an amateur geologist and archaeologist, he was active in many other fields, including social affairs, with publications entitled _De l'éducation du pauvre: quelques mots sur celle du riche_ (1842) and _De la femme dans l'état social, de son travail et de sa rémunération_ (1840), literary works, various essays on the creation and on the progression of life (1839-41), and on spontaneous generation (1862). In addition he established or promoted local literary and scientific societies in several towns of northern France, of which the grandly named, _Société Impériale d'Émulation d'Abbeville_ which he served as founder-president for many years, was the most prominent, and served as the principal vehicle for the promotion of his scientific and other views. However, he was never fully accepted into the national scientific establishment represented by the _Institut_ and the _Académie_ in the way that the Royal Society readily accepted into fellowship the equivalent British middle-class amateurs such as Prestwich and William Pengelly of Torquay. By 1863 Boucher de Perthes had been finding Palaeolithic flint hand-axes (_haches_) in the old upper terrace gravels of the Somme valley around Abbeville for more than thirty years but, as in Britain, little notice was taken of these finds until the late 1850s. Hugh Falconer is generally regarded as the first leading British geologist to investigate Boucher's finds seriously, during a visit to Abbeville in the autumn of 1858 when he'became satisfied that there was a great deal of fair presumptive evidence in favour of his [Boucher's] many speculations regarding the remote antiquity of these industrial objects and their association with animals now extinct'. (11) The first season of his work in Brixham Cave, Torbay, Devon, in 1858 had proved the contemporaneity of man and extinct large mammals in Britain. (12) Prestwich followed Falconer in April 1859, accompanied by John Evans, and confirmed Falconer's findings, seeing a worked flint still in situ at a depth of seventeen feet in undisturbed gravel, possibly the specimen shown in two original photographs in the Falconer Papers. (13)

In the spring of 1863 Lyell published his Antiquity of man in which the real contribution of Falconer (and of Prestwich and Evans) was almost completely ignored, (14) greatly distressing Falconer: 'I must keep my temper, for a serious onslaught. What a grumpy old Uncle you have got! Like an infuriated Toro - first having an encounter with Huxley - then goring Owen and his jackall - and then driving his horns into the porcelain ware of Sir C.L.! There will be a grand wash of crokery [sic] !' (15) In addition his health was failing and he was temporarily estranged from members of his family and consequently living alone in London. (16)

Falconer was touring France when, on 9 April 1863, Boucher de Perthes announced that at the end of March a workman from the gravel pit at Moulin-Quignon-les-Abbeville had brought him a flint implement with a fragment of bone containing a human molar, and on 28 March a human jaw had been found in the same terrace, thirty-three metres above the River Somme. (17) Prestwich and Evans were in Abbeville to investigate Boucher's claim on 11 April, (18) and on 13 April Falconer, who was in Paris, wrote to Prestwich:

> I have seen Lartet who has shown me a drawing of the lower jaw human (from Abbeville) and a section of the locality where found. There are certain questionable matters in the case which require to be cleared up. We are going to see it at Abbeville. Be therefore cautious in expressing any opinion till we meet . . . (19)

Falconer travelled to Abbeville the same day and wrote from there to Prestwich on 16 April:

> All right about the lower jaw, human. I have examined it severely and described it. It is not only fossil, but I believe of a race different from any known European race! However I say this with reserve as I have no recent jaws of the human race to compare with. I take with me a fragment of human jaw found at Menchecourt. Quatrefages came yesterday from Paris and carried off the Moulin-Quignon jaw . . . (20)

Two important points should be noted. First, Falconer had been convinced by the arguments that the jaw showed allegedly primitive features which he discussed in his letter to The Times published nine days later. Second, although Boucher's original account, as published in L'Abbevillois on 9 April, implied that the original 'detached molar' and the jaw found on 28 March were from the same site (Moulin-Quignon) Falconer's letter of 16 April attributes this to a quite different site (probably of a different age in the terrace sequence) at Menchecourt. Boucher had been collecting from the terrace gravels for more than twenty years but no human bones had been found: the extraordinary coincidence of two such finds in one week (probably from two different sites) ought to have raised more than a little suspicion, especially if Boucher de Perthes was offering 200 francs (fifteen weeks' pay for the local quarrymen) for the first human bone _in situ_ as has been alleged. (21)

W. B. Carpenter (a leading invertebrate zoologist and a Vice-President of the Royal Society) was back in London by 16 April after examining the jaw in Abbeville and spoke immediately to the Royal Society declaring the find to be authentic and quoting Falconer (without his permission) in support. (22) Falconer returned to London on 18 April and immediately began a detailed investigation of both the separate tooth that Boucher had let him bring to England and of his detailed drawings and measurements of the Moulin-Quignon jaw itself. After more than thirty years of brilliant taxonomic work, first in botany and more recently in vertebrate (especially mammalian) taxonomy, Falconer was too experienced to rely on the textbook drawings and descriptions of skull characteristics, and quickly turned to a friend who had a large collection of medieval and later human skulls from a churchyard at Clerkenwell, London. This population produced jaws showing all the allegedly 'primitive' characters of the Moulin-Quignon jaw (four allegedly 'Eskimo' features plus what was described as a 'marsupial-like' feature, sometimes found in Australian aborigines).

On sawing through the 'detached molar' he found that it was quite unfossilised and apparently comparatively modern, still retaining fresh gelatine. More seriously the tooth seemed to have been artificially coloured (as had the jaw which 'yielded a dirty white colour' when washed). Moreover, Prestwich and Evans had reported that _all_ of the flint '_haches_' allegedly associated with the human finds were modern forgeries. (23) Evans had already expressed his doubts to Boucher: 'Found M. Boucher de Perthes in a great state of excitement over a human jaw which had been found in a black seam at Moulin-Quignon which has been marvellously prolific lately . . . I doubt the whole affair and hurt poor old Boucher's feelings considerably.' (24) On 23 April Falconer wrote to Evans: 'I have sent a letter to the 'Times' - and I think - from the august auspices under which it has been sent they will give it insertion.' (25) This long letter was published in _The Times_ (which made room for it by dropping the daily Law Report) on 25 April and set out in detail the events of the previous four weeks. Falconer commented publicly on the sudden upsurge in the number of specimens of various kinds - especially spurious flint implements - being found: 'The number which turned out was marvellous, but the _terrassiers_ were handsomely paid for their findings, and the crop of flint-hatchets became in like degree luxuriant.' He then described the jaw, his comparative anatomy work on the sample of British jaws, and the character of the 'detached molar' which had 'proved to be quite recent; the section was white, glistening, full of gelatine and fresh-looking'. Falconer concluded:

> There was an end to the case. First the flint hatchets were pronounced by highly competent experts (Evans and Prestwich) to be spurious; secondly, the reputed fossil molar was proved to be recent; thirdly the reputed fossil jaw showed no character different from those that may be met with in the contents of a London churchyard. The inference which I draw from these facts is that a very clever imposition has been practised by the terrassiers of the Abbeville gravel pits - so cunningly clever that it could not have been surpassed by a committee of anthropologists enacting a practical joke . . . The break down in this spurious case in no wise affects the value of the real evidence, now well established, but it inculcates a grave lesson of caution. (26)

On the day this letter was published, Falconer wrote to Evans:

> Throughout, I contended that all the flint haches - twenty of them - that I took over from the Couche Noire were modern, and I never yielded the point. How those that tumbled out, when we were there, got in, I cannot say, nor can I explain the mysterious history of the antecedents of the jaw.
>
> But from its characters, which are irreconcileable with antiquity, or with it having lain long in a bed of iron or manganese, I deny that it can be older than the modern period. (27)

By this time Falconer must have received the two letters dated 19 April and 20 April 1863 that Boucher de Perthes subsequently published, (28) and which confirm that the 'detached molar' was from a different site at Menchecourt, from which additional specimens of animal bones had just been obtained in the presence of Nicholas Brady (son of Sir A. Brady, a prominent member of the Geological Society) who was bringing the additional evidence back to England for Falconer to see.

On about 27 April Falconer received a further extremely long letter from Boucher dated 26 April, also published: 'Dear Doctor, The rumours in England about the falsification of flints at Abbeville are contrary not only to truth but also to reason . . .' He vehemently denied Falconer's allegations against the quarry workmen: 'No, a thousand times no, that is impossible.' (29) After detailing the distinguished support that he was receiving in France, he claimed that the workmen had rejected the jaw and teeth as worthless, and that 'without Mme. Dufour they would never have been found'. He concluded with an extraordinary story (which, incidentally, shows the degree of popular interest in the discovery):

> Meanwhile, our spiritualists in Paris are calling up the fossil man of Abbeville. An eye witness writes to me, with all seriousness, that during a spiritualist seance which took place last Thursday, 23rd April, the fossil man declared that the disaster which caused his death, twenty thousand years ago, crushed him between two stones before depositing his jaw where I found it, and that his name was Joé etc. . . . The deceased antediluvian did not seem to bear a grudge against me: that is generous, for in my book on Masks, I wrote an article against spirits and spiritualists . . . (30)

Falconer was greatly disturbed by events:

> I am absolutely alarmed to peruse Boucher de Perthes letters - first from their acreage and next from the desolation of feeling they display. But the most serious part of the Case is that Quatrefages gave a communication to the French Academy on Monday last backing up Boucher de Perthes' memoir and citing me several times. I wrote to Lartet on Tuesday and Wednesday, to get him to put him and Quatrefages on their guard - but too late. Further by a letter received this morning from Lartet I learn that Delesse had examined one of the flints and considered the brown hematite spots to be veritable marks of antiquity! and he gave the same opinion regarding the jaw! But Lartet has his suspicions - in unison with ours. (31)

However, the publication of the letter in The Times prompted Falconer's niece, Mrs Grace McCall, to write to him, and he replied to her in a letter, marked 'Private', on 28 April:

> Thanks for your affectionate note. I know that my letter in the 'Times' would give you all some gratification . . . But there is no rose without a spine. It has put me in conflict with the French Academy. Quatrefages gave a communication on Monday last to the Institut - affirming the authenticity of the specimen, mainly on my first mistaken impressions - for I was deceived at the outset - by first appearances, but not . . . when I went seriously into the evidence. The truth has been spoken out and I spoke it. Fancy what discredit would have been thrown on the subject had the expose been made by the enemy - such as Soapy Sam [Bishop Samuel Wilberforce]! We would have been regarded as simpletons open to be practiced upon by the flimsiest imposition: and the whole subject would have been put back quarter of a century. I confronted it - and I find the course has been good . . . Poor dear Boucher is desole. Perfide Albion! I do believe he has turned his face to the wall, but I hope to curse God and die. Oh

> that you were here as a buffer to toil through the acreage of his compositions, and to sustain the pang of his wounded feelings! . . . (32)

The next day he wrote to John Evans:

> Our dear friend Boucher is beside himself. I have had to pay a shilling extra postage this morning on a letter from Abbeville - all written documents. Further - on Monday week Quatrefages gave a communication to the Academy of Sciences, affirming the authenticity of the jaw - and chiefly from the strength of my asserted convictions! Lyell-Boucher-Dr. Carpenter-the French Academy-and Mr. Alfred Tylor - all upon me at the same moment. What shall perfide Albion do? Could you contrive to bury me in the Biddenham Pit? and what would suit you to go to Bedford? I do feel as if a slight dose of burying alive would be rather a wholesome diversion . . . By a letter from Lartet Delesse after careful examination of Quatrefages hatchets believes them to be ancient: and he finds the same marks of authenticity - in the jaw. I have had letters from Quatrefages and Lartet. a dreadful position altogether. (33)

Boucher continued to write despairing letters to Falconer, but these were presumably too emotional (or repetitive) for him to reproduce in his 1864 volume. Falconer gave the latest news in a letter to Evans on 5 May 1863:

> Nicholas Brady has made his appearance with a fresh consignment of 8 flint implements from 'Moulin Quignon'. At once examining them I pronounced 6 to be modern fabrications: two being veritable ones that may have come out of anywhere. Prestwich examined them two hours afterwards and gave precisely the same verdict 6 spurious 2 real. They were sent over by Boucher - or rather with his assent as 'pieces justifications' to confound us.
>
> I am sorry to tell you that Boucher is very unwell. The Brady's report seriously of him. He has taken the matter sorely to heart. His notes to me are unreadable from their severity. They are peppered over with daggers 'vous m'avez donné un coup funeste' [You have struck me a mortal blow.] etc.
>
> I wish you would send me your false hache from 'Moulin Quignon' and any true ones. I shall be obliged to go over to Paris this week to see to the matter or rather confront it. Quatrefages, Delesse and Desnoyer,according to Lartet maintain the authenticity of the haches found by Quatrefages (one of them at least) and of the jaw . . . (34)

As Charles Murchison put it a few years later: 'Men of science in France and England were suddenly placed at direct issue on a grave and important point of great general interest.'(35) The leading French workers proposed that Boucher's finds and claims should be examined dispassionately in a quasi-judicial manner under the auspices of the _Institut_, _Académie des Sciences_ and Royal Society by a joint Anglo-French commission of inquiry. Later the same day Falconer wrote a second letter to Evans:

> Since writing you Dr. Carpenter called with a _cartel_ from M. Quatrefages. I am invited to go over and do battle in the matter of the Moulin Quignon jaw.
>
> Carpenter _pro_ and I _con_ start by the mail train on Friday next the 8th for Paris and to remain Saturday and Sunday. We meet Quatrefages, Lartet, Delesse and Desnoyer. You and Prestwich are invited also. one of you _must_ come. I do not think Prestwich _can_: you therefore must. Arrange it between you.
>
> Any how send me your forged 'Moulin Quignon' hache - and any help you can give me. Reply by return and oblige. (36)

The same day he also wrote to Prestwich:

> Make your arrangements instanter. Dr. Carpenter has called on me with a formal _cartel_ from Quatrefages, challenging me, you and Evans to go over to Paris, and do battle about the Moulin-Quignon human jaw. I have written to Lartet accepting. Carpenter as 'avvocato di Diavolo' i.e. _pro_ and I _con_ start by the mail train of Friday next 8th. for Paris. Either _you_ or _Evans must_ come. He _cannot_ - you can. Get ready, Oh Gravel sifter! and send me anyhow all your forged Moulin-Quignon haches. Try and get Alfred Tylor to deliver up his one. (37)

The following day, 6 May, Falconer wrote again to Evans, pressing him again to join the commission and dismissing a suggestion that 'G.B.' (presumably George Busk) would be an adequate replacement that Evans had presumably made:

> Prestwich has sternly decided against going, notwithstanding my moving implorations - and unless you come Albion is consigned to mortification and defeat. Let me have a reply quickly - saying that you are coming. We leave on Friday by the mail train . . . Your _deference to_ and _laudation of_ G.B. as so competent and able a judge are not to be endorsed. I give you up! (38)

The joint Anglo-French commission opened on Saturday, 9 May 1863, and Falconer outlined the arrangements and the first day's deliberations in a letter from Paris to Grace McCall the following day:

> Yesterday at 11 I called at the Mohl's saw him but he is laid up with fever then to Lartet and the Jardin des Plantes. We got Milne-Edwards who is half English to preside as moderator and formed our deliberative bench thus.
>
> President and Modtr. M. Milne Edwards
>
> | French assessors | English assessors |
> |---|---|
> | 1. Quatrefages | 1. Falconer |
> | 2. Lartet | 2. Dr. Carpenter |
> | 3. Delesse | 3. Busk |
> | 4. 'Desnoyer | |
>
> Secretary et Redacteur du procès verbal. Delesse
> other French Savants present
>
> 1. M. Buteux) to represent M. Boucher de Perthes
> 2. L'Abbé Bourgeot)
> 3. M. Gaudry
> 4. M. Milne Edwards fils
>
> We have been met in the best possible spirit and are going thoroughly into the Case. We sat till 5 p.m. yesterday, it was the outlying points we went into - in order to establish certain facts as a basis for judgement on the real points of the case. I do not think that we have lost any ground and I have confidence from the present aspect of the matter, that we shall establish our case. But I am open to Conviction and will give a true and honest verdict to the best of my convictions. That letter to the Times has been like a bomb shell. (39)

The members of the commission covered a wide range of scientific expertise. The president, Henri Milne-Edwards (1800-1885), was the son of an English colonist from Jamaica. He was born in Bruges although he opted for French citizenship when Belgium became independent. He was a good example of the kind of broadly based 'environmentalist' discussed by Ludmilla Jordanova in her contribution to this volume: medically trained in Paris, he worked in zoology and his first major appointment was as professor of hygiene and natural history in the *École Centrale des Arts et Manufactures*, four years after being awarded the *Académie's* prize in experimental physiology in 1828. He subsequently established an international reputation in the physiology and comparative morphology of marine invertebrates during his tenure of the post of professor of entomology (which covered all arthropods) in the *Jardin des Plantes*. At the time of the Moulin-Quignon affair he had fairly recently transferred to the chair of mammalogy at the Muséum in the Jardin des Plantes following the death of Geoffroy Saint-Hilaire.

The other French participants covered a wide range of scientific interests, the outstanding figures being Quatrefages, Lartet and Delesse. J. L. Quatrefages (1810-1892) first studied mathematics, physics and chemistry at Tournon and then medicine at Strasbourg. He was closely associated with Milne-Edwards in studies of the invertebrates and other marine biological work from 1840 to 1855 when he was appointed (with the support of Milne-Edwards) to the chair in the natural history of man in the Muséum d'Histoire Naturelle, although his work on invertebrates continued. As an anthropologist he was an anti-Darwinian, concentrating on comparative human morphology and on the unity of the human species and the non-simian origin of man. Edward Lartet (1801-1871) qualified as a lawyer and then managed his family estates in the Gers district of south-east France where he became interested in palaeontology, working on the fauna of the local Tertiary and Quaternary. In 1836 he discovered the first specimen of *Pliopithecus*, the first fossil anthropoid ape to be recognised, and he then turned to the search for fossil human remains. Moving to Toulouse he carried out many excavations on prehistoric sites of the Perigord including (with the Englishman, Henry Christy) the classic Upper Palaeolothic sites of La Madeleine, Le Moustier and Les Eyzies in 1863. He was elected the President of the first International Congress of Archaeology and Prehistoric Anthropology, held in 1867. His only academic appointment was as professor of palaeontology in the Muséum d'Histoire Naturelle in 1869, and he soon had to give this up because of ill-health. A. E. Delesse (1817-1881) was educated at the École Polytechnique and joined Elie de Beaumont and Dufrenoy at the École des Mines where he worked first on the preparation of a catalogue of collections of geological maps, travelling extensively in Europe, studying geology at the same time. After a period in the chair of mineralogy at Besançon he returned to Paris, first to the Sorbonne, then the École Normale Supérieure and, in 1864, to the École des Mines to teach a new course in agriculture, drainage and irrigation. Subsequent work included major studies of hydrology and the geology of the ocean floor.

The British 'jurors' at the enquiry were also a diverse group in terms of their interests and experience. Apart from Falconer, the best known was W. B. Carpenter (1813-1885). He had qualified in medicine at University College, London, and continued his studies at the Edinburgh Medical School. He first made his mark in the field of physiology. At the age of thirty-one he became Fullerian Professor of Physiology at the Royal Institution and was also elected a Fellow of the Royal Society. He soon held a number of other posts in London, including the forensic medicine chair at University College and the Swiney lectureship in geology at the British Museum, and finally served as Registrar of the University of London from 1859 to 1876. In addition he was a great all-round naturalist, making major contributions in the fields of marine zoology, especially studies of the foraminifera and crinoidea. Through a range of more popular scientific works and cyclopaedia contributions, and his attacks on contemporary pseudo-scientific cults such as phrenology, spiritualism and mesmerism, he developed a wide popular following, particularly amongst more progressive amateur naturalists. An active Unitarian, he did not entirely accept Darwin's views on evolution, arguing that the theory of natural selection did not destroy traditional theistic arguments of Design in creation. George Busk (1807-1886) also qualified in medicine (as an articled student of the College of Surgeons) and served as a surgeon in the Navy from 1832 to 1855. On his retirement he devoted himself to his natural history interests. (He had been elected to the Linnean Society in 1846 and an FRS in 1850.) He was primarily interested in microscopy (he was a founder of the Microscopical Society in 1839 and its President in 1848 and 1849) and, apart from his recent election to the Council of the Royal College of Surgeons and his following as a popular editor of the Natural History Review, appears to have had no relevant experience at the time of the Moulin-Quignon commission and one can understand Falconer's concern at the nomination of such a 'lightweight' as a British representative. However, Busk subsequently travelled with Falconer to investigate the Gibraltar fossil man finds, and developed an interest in comparative osteology, describing the animal bones in the Brixham Cave excavation report. He was a founder council member of the Anthropological Institute, and its President in 1873 and 1874.

Subsequently the British contingent was augmented with the arrival of Joseph Prestwich (1812-1896), who was at the time still running his family's wine merchant business in the City of London. After a year at University College, London, where he was particularly attracted to science, he joined his father's business at the age of eighteen, and continued his studies privately in his own time. Concentrating on geology, he soon attracted notice as a promising geologist with two papers on the geology of Coalbrookdale read to the Geological Society, but he subsequently specialised in the Tertiary geology of the London area and of southern England, at the same time carrying out pioneering work in hydrology, which resulted in a memorable paper on the water supply of the London area (1851) and his subsequent election to the Royal Society. He was one of the pioneers of the study of fossil man, participating in the Brixham Cave excavation in 1858 and giving a classic paper to the Royal Society in May 1859 supporting the claim of Boucher de Perthes to have found humanly-worked tools in association with extinct animals, such as mammoths, in the ancient terraces of the Somme. He became President of the Geological Society in 1870, and in the same year married Grace McCall, Falconer's niece and confidante. In 1874, two years after his retirement from business and at the age of sixty-two, he accepted the chair of geology and mineralogy at Oxford (originally established by Buckland) on the death of John Phillips. (Since he had no academic qualifications whatsoever, Prestwich was hastily awarded an MA degree by the University on taking up his appointment.) His great scientific strength lay in careful and systematic stratigraphical research, recording and collecting. Philosophically, Prestwich adopted an actualist position - accepting much of Lyell's uniformitarian arguments, certainly in terms of process, but arguing against a strict uniformity of force or effect, in favour of a position that accepted larger scale natural forces from time to time. In his later years he spent much of his time seeking evidence for a large-scale early Quaternary flood which, he postulated, had covered much of England and northern Europe: evidence which has again been the subject of some geological controversy in the mid-1970s.

The first day of the Anglo-French commission's deliberations, together with the second full day session, was devoted almost entirely to a detailed examination of the recently found flint implements from Moulin-Quignon and comparing them with established specimens from other sites around Abbeville, and on the criteria that should be used for distinguishing forgeries. (40) Falconer wrote to Grace about the second day on 11 May, and told her that Prestwich had changed his mind and had unexpectedly arrived:

> I wrote you a short note yesterday just as I was starting for the Commission. Mr. Prestwich, to my great relief made his appearance suddenly and quite unexpectedly, a powerful reinforcement to our array. We sat from 11 am to 6½ pm. fighting every part of the evidence sternly on the outlying points. Today is to be our final sitting - and then we go to Abbeville. We go into the characters of the jaw today. I shall be obliged to return to Paris for a few days to do some work with M. Lartet, and then I shall return straight to London. Last night after the Seance was over we all dined 16 or 18 at M. Milne-Edwards. We have not lost ground so far, and my impression is that we shall hold our own - but that remains to be seen . . . The affair of Abbeville is being discussed in all the French papers . . . M. Mohl is suffering from feverish attacks. I have not seen Madame. In fact I have been unable to do anything - or go anywhere on account of this terrible affair of the fossil man of Abbeville . . . (41)

The third session opened with an examination of the earliest find, the 'detached molar' from Menchecourt that Falconer had studied in London. This was rapidly dismissed on grounds that are, in the literal sense, quite incredible:

> But on this subject M. de Quatrefages remarked that there might be some uncertainty on the siting of this piece, because M. Boucher de Perthes had several human teeth, found in the same formation, on different spots around Abbeville, and that the savant, having taken all the objects out of their respective boxes at the same time to show them to M. Falconer feared he did not put each article back in its place, which could have led to an error in the labelling of the boxes. (42)

The idea that Boucher could have 'lost' this unique first specimen - which he had been searching for over a period of more than thirty years - or put it in the wrong box, and consequently given Falconer the wrong specimen for investigation during his April visit, is too far-fetched for words and, coupled with the extraordinary coincidence of two separate finds in a single week after such a long period of collecting, must have deeply disturbed those who were already sceptical. Moreover, whereas Falconer had merely accused the quarrymen of fraud, Quatrefages' clumsy excuses might be interpreted as implicating Boucher de Perthes himself: at the very least they implied quite extraordinary carelessness on his part.

The commission then turned to the Moulin-Quignon jaw itself, and at Falconer's insistence, in the presence of all participants, Busk sawed across the jaw immediately in front of the second molar. The staining on the outside of the jaw was easily washed off, the interior of the jaw produced traces of a fine grey sand - quite different from the Moulin-Quignon matrix. The sawn bone and tooth was relatively fresh, with the dentine of the root of the tooth 'perfectly white, full of gelatine, and in no respects different in appearance from that of a recent tooth'. (43) Falconer also produced the collection of jaws from the London churchyard mentioned in his letter to The Times for comparison with the Moulin-Quignon jaw. (44)

Falconer again described the previous day's proceedings to Grace in a letter written at 5 am on 12 May:

> This morning I am about to start with the Commission for Abbeville. Yesterday we went into the evidence of the jaw itself - and I am happy to tell you that not one character was made out to our conviction justifying us to consider the jaw as an authentic fossil.
>
> Prestwich, Busk and myself held to this and some of the French advocates were utterly shaken in their faith. For we detected [the English did] in the interior of the jaw, irresistible evidence - of an anterior prolonged sepulture: i.e. that the jaw had undergone one burial before it found its way into the 'Black-Bed' of Moulin Quignon. You will at once perceive to what this necessarily leads. But as the Commission has not drawn up its

report we must express ourselves with reserve. Virtually the case is at an end. My friend Lartet, a profound palaeontologist admits the evidence of imposture.

You will have no conception of the interest excited by the subject in Paris. It is known that we are here, and near the Jardin des Plantes we are pointed out as les "audaces anglais" [sic] - who question etc. The French made the most determined stand up fight throughout, and combatted every point of evidence, even the plainest and every position taken up by us. But, to no avail.

The Case has become a "Cause Celebre". Do not let what I have told you get published, except that you are quite authorised to say, that the English[illegible ? savants] after going fully into the case saw no evidence to warrant the belief that the jaw was an authentic fossil or to shake their original opinion. (45)

After spending two days in Abbeville (without Prestwich who had returned to England) the commission concluded its proceedings. Milne-Edwards prepared an immediate report for the Académie des Sciences in addition to the full procès verbal that was being prepared by Delesse. Falconer back in Paris, wrote a letter marked 'Private' to Grace on 14 May:

I returned from Abbeville last night. The Commission is at an end. After a determined tussle of five days, the case has gone against us. But the evidence was so strong in our favour - that till the last day it seemed as if the day was ours. The French Secretary was even discussing with me the terms in which the non-authenticity of the jaw could be most easily let down! I told you about getting whitish grey sand inside, when the jaw was sawn across, a very strong looking proof of previous sepulture. The other appearances were all in Conformity. But in the long run matters were against us and I am glad that they did, as truth alone was our object. I enclose a copy of my final conclusion. Busks was in other words - but practically to the same effect, we believe the jaw to have been found naturally where it was got, and think that there was not fraud in the Case: but that the jaw is not of any great antiquity. Prestwich's convictions went entirely with the French. He felt crushed by the evidence of the last day. I am very much done up. One night - the last - I did not sleep a wink. But I made up for it last night. We have parted from the French on the best of terms - and in the best spirit. They admit our fairness, the absence of any spirit of advocacy or partisanship - and that we had the very best grounds for our suspicions. Therefore, my good Deductive, do not be cast down although your Uncle has been concerned as a loser in a "Cause Celebre". When the evidence is given and the whole tale is told, I question if any judge or jury in England would admit the

> authenticity of the jaw. Yet I do, and I will do my best to support it. (46)

With this letter was enclosed a copy of a certificate signed by Falconer, marked '<u>not</u> <u>to</u> <u>be</u> <u>used</u>', and dated 'Abbeville 13th May 1863': 'I am of opinion that the finding of the human jaw at Moulin-Quignon is authentic: but that the characters which it presents taken in connexion with the Condition under which it lay, are not consistent with the said jaw being of any very great antiquity.' (47) The next day (15 May) Falconer amplified his views in a further letter marked '<u>Private</u>' to Grace:

> I am just about to leave for London by the mail train . . . Do not be cast down. I still maintain that the jaw is not of remote antiquity - and I will tell such a fine tale to the British public as will make John Bull furious that I admitted <u>authenticity</u> of finding to any extent. But the honor of an amiable old gentleman and the susceptibility of a fiery nation were concerned. Fancy, the tooth <u>in</u> <u>the</u> <u>jaw</u> was nearly as fresh - as the detached tooth of "<u>Times</u>" notoriety! All this for your own discreet ear. Do not blab or say anything, lest I should be compromised. Never, I believe, in the history of Science, was such a perplexed and incompatible case heard of . . . Keep this note - or return it to me as a voucher of Lartet's opinion. (48)

Clearly, Falconer was far from being satisfied by the outcome of the proceedings, but this is not reflected in the report of Milne-Edwards to the <u>Académie</u> three days later:

> All the members of this gathering of friends embraced the same opinion. Dismissing any idea of fraud, they recognised, in the most candid way, that there no longer seemed to be any reason to call into question the authenticity of the discovery made by M. Boucher de Perthes of a human jaw in the lower part of the large deposit of gravel, clay and pebbles of the Moulin-Quignon quarry. (49)

The formal conclusions of the commission were as follows (all the French members plus Prestwich adopting all four clauses):

> 1. The jaw in question was not fraudulently introduced into the Moulin-Quignon quarry; it was previously in the place where M. Boucher de Perthes found it on 28th March last. This conclusion was unanimously adopted.
> 2. Everything suggests that the deposition of this jaw was contemporary with that of the pebbles and other materials which make up the clay-gravel mass called the 'Black Bed' which rests immediately on the chalk . . . Messrs Falconer and Busk reserve their judgement . . .

> 3. The flaked flints in the form of axes, which were presented to the meeting as having been found around the same time in the lower parts of the Moulin-Quignon quarries are mostly, if not all, quite authentic . . . Mr Falconer reserves his judgement.
> 4. There is no sufficient reason to doubt the contemporaneity of the deposit of flaked flints with the jaw found in the Black Bed . . . Messrs Falconer and Busk . . . wish to reserve their judgement.

Falconer entered a note of reservation in the terms stated in his letter to Grace of 14 May (see above), while Busk added the following:

> Mr Busk desires to add, that although he is of opinion judging from the <u>external</u> condition of the jaw, and from other considerations of a more circumstantial nature, that there is no longer reason to doubt that the jaw was found in the situation and under the conditions reported by M. Boucher de Perthes, nevertheless it appears to him that the <u>internal</u> condition of the bone is wholely irreconcileable with an antiquity equal to that assigned to the deposits in which it was found. (50)

In his own account of the proceedings, written in 1863 but not published in his lifetime, Falconer explained clause by clause the exact meaning of his note of reservation:

> The first clause of my verdict 'that the finding of the human jaw at Moulin-Quignon is authentic' was intended to absolve the workmen from the imputation of having fraudulently introduced the bone into the bed, when no direct proof could be adduced to support it; while by the second clause it was meant to express my opinion that the bone was not of fossil antiquity, i.e. not reaching further back than some date in the modern period . . . a wider meaning appears to have been attached to my use of the term 'authenticity of finding in the communications . . . by M. Milne-Edwards and M. Quatrefages on the 18th of May' than I intended the words to convey. They did not include an admission of the 'authenticity of the jaw' as a true fossil bone . . . from first to last I entertained an adverse opinion on this head . . . I now believe that I committed an error of judgement in not reserving my opinion on <u>all</u> the moot points, instead of reserving it upon three only. I must bear the blame which this admission carried with it, considering how strong my negative convictions, founded upon the intrinsic evidence, were. I have since carefully reviewed the facts and opinions set forth in the 'proces-verbaux', and submitted to the closest examination the numerous suspected flint <u>haches</u>, yielded by the 'black seam' and ferruginous gravel of Moulin-Quignon immediately before the Conference and while it sat there; and the result is an irresistible impression, that there is some mysterious

complication in the case which remains to be solved. (51)

Falconer wrote to The Times for a second time, and this was published on 21 May 1863. Referring to his April letter denouncing the Moulin-Quignon jaw and associated finds, he withdrew his allegations of fraud by the quarrymen, but was otherwise very cautious:

> [The jaw] . . . has since undergone an investigation at Paris and Abbeville by a joint commission of French and English men of science, throughout which it maintained the same perplexed and contradictory character, not to be surpassed, in some respects, at least, by any *cause célèbre* on record. But I am happy to say that upon one point, which it was of the last importance to clear up, the commission, French and English, were unanimous - namely that the discovery of the remarkable human relic *in situ*, in the gravel pit of Moulin-Quignon, was authentic, and that no imposition had been practiced by the workmen in the case. As an inference to the contrary on the part of myself and my scientific friends was distinctly expressed in my former letter, I am desirous that there should not be the slightest reserve in withdrawing it. What now remains to be established is the precise age of the relic. This part of the case is still involved in obscurity, and so beset with contradictory and, apparently, incompatible evidence that its satisfactory solution is at the present moment of the utmost difficulty . . .

He continued by giving an account of the events of April and of the meetings of the commission in Paris at the end of which:

> So far, no point had been established to shake the confidence of the English members on the soundness of their doubts . . . Two, at least, of the four French members frankly and openly admitted the effect which the evidence yielded by the section . . . of the jaw had produced on their views; and had the inquisition been carried no further it is probable that the result would have been a verdict of 'not proven' . . . (52)

However, during the subsequent visit to Abbeville they had seen a most convincing demonstration in the Moulin-Quignon gravel pit where workmen digging in apparently undisturbed gravels, under the supervision of members of the commission, produced five *haches*, four of which were very similar to the allegedly modern forgeries. This demonstration greatly demoralised the 'opposition' on the commission, and all signed the formal conclusions reproduced above subject to the reservations detailed by Milne-Edwards. Despite this Falconer was still convinced that 'the question of the antiquity of the relic still remains to be determined'. He added, in a postscript, that Carpenter 'had been sufficiently impressed . . . to have come to doubt his original belief in the authenticity of the *jaw*'. (53)

The day that this letter was published, Falconer wrote to John Evans expressing his growing doubts, and asking Evans to intervene:

> The fossil aspects of the jaw have been exhausted. All depends now upon the proof of where it was found - and of what age the deposit. But it is otherwise as regards the haches. I believe that they are determinable by their inherent character - and that all the 'black seam' lot are of a comparatively modern character. I have again gone carefully over the evidence of the whole lot which I took to Paris and with the disastrous experience of my last visit to Abbeville find my convictions hold to the determination that they are not of the old type - but of a more modern character. You are the Achilles in this walk of the Strife and you are bound to come forward in some shape or other - to speak to the faith that is in you and upon which others have relied. I have an excellent lot of material but in a few days they will all be dispersed - among the various owners. You have seen my letter to the 'Times' - where I have held fast - where others have backslidden. You are bound to come forward. (54)

The following day, 22 May, Falconer wrote again to Evans saying that 'the flint haches maintain their modern aspect', and continuing:

> But a good solid letter would come with more effect from you as you have not been compromised by any change of opinion - and you could therefore lay it on handsomely on these villanous personations of antiquity. You must be quick about it - if you do anything. Further Prestwich says you and he think of going again to Abbeville. In that case you must attack the cliff of the Gravière horizontally from the top downwards along an extent of 5 feet wide by 30 feet long taking up the deposits like sheets of paper as we did in the Brixham Cave. In that case - nothing could escape your observation. It would be a crucial test. I have no faith in the vertical mode of attacking the cliff. Nothing is seen in situ - or only in a sham fashion. Din this well into Prestwich's ears. I have not yet forgiven him. (55)

On 23 May, Falconer wrote again to Evans again asking him to attack the authenticity of the haches regardless of the possible effect on Falconer's own reputation:

> Lay on therefore - and write without mercy - the haches me and Prestwich - especially - arrange with P. about coming out with an article and do not lose time about it for the Athenaeum or Times. I shall be truly glad to see you when you can come - for really I am sorely afflicted at the result arrived at. (56)

Falconer was by now convinced that the demonstration for the commission at Moulin-Quignon was the most spectacular fraud of all, and on 25 May he wrote to Prestwich:

> Remember, throughout, I denied the authenticity of all the haches, shown in Paris from the 'black bed' Brady's, your own, mine, Quatrefage's. To that opinion I still adhere. How those tumbled out before the Commission, or how they got in I cannot say. I insist merely that they are not *ancient*. Further I do not admit the *antiquity* of the jaw. The fresh characters etc. are utterly irreconcilable with that. I do not believe that it goes further back than some date in the modern period. (57)

In his next letter to Evans, on 29 May, Falconer was even more explicit about his belief that the whole 'Affaire Moulin-Quignon' was a major fraud, and doubts were now being expressed in France:

> I am more convinced than ever that there is some terrible mystification in the Moulin Quignon case. Elie de Beaumont is denouncing it (at two meetings of the Academy) as as modern as a turf-bed and likely to yield Gallo-Roman Coins. Prestwich yielded too readily. But I have a strange notion that a £10 note, well applied, might elicit some secret knowledge that might dispel the obscurity. I wish you and Lubbock and Austen would try to see if there is the means of applying this very effective means of investigation - in certain cases. I will join with you in a contribution . . . P.S. Prestwich is *too good* to have such a hint even broached to him. I therefore apply to you. (58)

In June, Falconer received Delesse's draft of the *procès verbaux* and spent a great deal of time working on this, adding many detailed notes, and preparing an English translation for the *Natural History Review*. (59) Falconer's retraction of his allegations of fraud greatly improved relations with France, and Boucher de Perthes wrote to him on 4 June following Falconer's (apparently unexpected) unanimous election as an honorary Association Member of the *Abbeville Societé d'Émulation*, of which Boucher was President. Boucher commented:

> *The case of the jaw*, as the learned and witty Dr. Carpenter calls it has given you much anxiety, as it has to me too; but I comfort myself because it has not affected your friendship for me nor mine for you . . . It is pitiable to see passion expended on such matters, which should remain in the peaceful realm of science; but many people, both in France and England, have turned it into a partisan matter: hence all the anger. (60)

By this time, however, Evans had found clear evidence that one Moulin-Quignon specimen was a blatant forgery, with clear marks that it had been manufactured with the aid of metal tools. Evans summarised this in his notebook on 31 May:

Vasseur picking at the foot of the bank, in the presence of Lubbock, Flower, and J. E. dislodged from the face of the couche noire from under a few inches of talus a hatche of oval form which was carefully picked up without touching its surface. On examination the surface, which was black and showed a slightly metallic lustre, showed the marks of fingers with which it has been smeared with the colouring matter. Vasseur having found the implement soon after gave up the search. (61)

Further investigations by British workers followed in June and on 20 June Falconer wrote to Evans:

I have now got Keeping's letter addressed to Prestwich with his final report. He reports all his findings as being 'plants'
No. 1 - 3 June - a 'plant' well laid
2 found in his absence
3 - 4 June @ 6 inches above the chalk gravel disturbed from a foot round - to his own conviction in working - and got the specimen in the centre of the plug.
No. 4 - 4th June - found by himself - after returning from dinner - [illegible] ground disturbed a 'plant' wants Evans to examine the smear.
No. 5 - 5 June: had previously observed a crack - the crack then filled up - (next morning) worked on it discovered the flint and the crack reappeared behind it! a 'plantissime'
No. 6 - 5 June found in his absence
No. 7 - 6th June - a 'plant' - found by himself
Concluded: 'I have every reason to believe that all the specimens I have brought from Moulin-Quignon were placed there on purpose for me to find.'

I do hope and trust that you will publish Keeping's experience. It is too much to expect it of Prestwich who is very unhappy at the facility with which he leapt into the arms of credulity and delusion. I am in an equally bad way - but it relieves me to see your neighbour overcome. (62)

The next day Falconer commented to Prestwich: 'Many thanks for the perusal of Keeping's sensible report. The grounds of suspicion are thick as autumnal leaves in Vallombrosa.' (63) Evans reported Keeping's findings in the Athenaeum, and all the British workers, including Prestwich, had withdrawn their support from the Moulin-Quignon 'discoveries' by the time the procès verbaux was actually published at the beginning of July, as Falconer made clear in a letter to John Evans dated 10 July:

I have just got back from Edinboro'. I saw your thunderer in the

> Athenaeum well done Keeping. Poor dear Mr. Prestwich will make all the world believe that the notes of his opinion given to the Conference in the abstract of his letter are his current views! and nobody could hit the spurious haches harder than he does. (64)

Boucher de Perthes devoted the greater part of the third volume of his *Antiquités*, published in 1864, to a defence of Moulin-Quignon (which by that time was supposed to have produced over one hundred human fossil remains) with reprints of various papers, correspondence etc. (65) However, within two or three decades, even in France, there was no longer any enthusiasm for the Moulin-Quignon finds, and it was suggested that the jaw had originally come from a Neolithic (or possibly Iron Age) burial in the Abbeville region that (after the surface had been artificially coloured) had been fraudulently 'planted' at Moulin-Quignon. (66) Oakley indicates that modern tests confirm that the jaw was sub-fossil (not completely fresh) and has suggested a Neolithic date. (67)

It is not the purpose of this paper to speculate on who might have been responsible for the series of frauds, but there is no doubt that this was quite a sophisticated venture, calling for something close to Falconer's hypothetical 'committee of anthropologists enacting a practical joke' rather than the humble '*terrassiers*' indicted in his first letter to *The Times*.

Aspects of the deception certainly worried Falconer and the other British members of the commission very much, and the lame excuse suggested by Quatrefages in respect of the Menchecourt 'detached molar' (and the apparent attempt in some quarters to pass this off as a Moulin-Quignon specimen) must have led them to wonder whether Boucher himself, or at least someone very close to him, was implicated. Falconer in particular paid generous tribute to the outstanding importance of Boucher's genuine finds in the Somme terraces, and there is a hint in his remark about 'Soapy Sam' Wilberforce (68) quoted above, that he might have considered the possibility that the fraud was a deliberate attempt to discredit Boucher de Perthes and his work on the antiquity of man.

The quasi-legal response from the French scientific community to the (largely if not wholly) British allegation of outright fraud was an interesting and very unusual one. Some English commentators (especially Charles Murchison) and the president of the Anglo-French commission suggest that at least part of the reason for this approach was to preserve and foster relations between the scientific communities of the two countries, and a somewhat similar approach was adopted four years later in investigating forged documents purporting to show that Pascal had priority in terms of scientific discovery over Newton. In the case of Moulin-Quignon however the procedure was, in relation to this interpretation of the objectives of the

Institut and _Académie_, a failure. In part at least this failure was due to the commission drawing back at almost the last moment (during the visit to Abbeville) after, according to Falconer's private letter of 14 May, even the Secretary (Delesse) was discussing the form of words in which the authenticity of the jaw could be rejected with the minimum public damage to Boucher's sensibilities and reputation. Clearly the visit of the commission to Abbeville (minus Prestwich who had to return to London) is the key to the failure of the commission. Possibly Boucher's desolation in the face of the allegations convinced the members of Boucher's personal innocence in the matter and they therefore did not pursue the (much more likely) alternative that someone other than Boucher was the local forger. The other, more important, factor was the very spectacular demonstration of the 'discovery' of items of the kind condemned by Britain's researchers, Prestwich, Falconer and John Evans, from apparently undisturbed ground, in the presence of the members of the commission.

The principal participants in the work of the commission covered a wide range of specialisations, mainly developed from an initial medical training. However most had little current experience or standing within the field of mammalian palaeontology or archaeology: Falconer was the only vertebrate palaeontologist and (of those who attended all the sessions) Lartet was the only one with a claim to expertise within archaeology. It seems clear that the main issue debated by the commission was not the authenticity of the jaw and the detached tooth as such but rather the character and mode of disposition of the disputed _haches_. There was only limited discussion of the anatomical or other features of the Moulin-Quignon jaw itself, and the majority seem to have given little attention to the two most damning pieces of evidence: the indication of artificial colouring and the evidence from the matrix and inclusions that the jaw could not possibly have come from the 'black band'. Also, the commission would not wait for proper chemical analyses of the bone and matrix, despite Falconer's insistence on the necessity of this. Moreover, the very first 'find' - the Menchecourt 'detached molar' - was simply withdrawn on the grounds that no one was sure that the specimen given to Falconer by Boucher less than three weeks after its 'discovery' was the correct tooth! Finally, when the commission inspected Moulin-Quignon, no proper excavation (using the techniques developed by Falconer, Prestwich and Pengelly in Brixham Cave from 1858 to 1860) was carried out. Instead a large gang of workmen excavated back the vertical quarry face and obligingly 'found' four fake _haches_ and one genuine one (probably imported from a different site) in the presence of members of the commission.

Even those who were originally quite convinced were soon overwhelmed by the improbability of the spate of human bones and flint implements from Moulin-Quignon and nearby pits in gravels of different geological ages, especially since the same implement types were being 'found' in deposits of very different ages, as was noted by Prestwich in October 1863. (69) Boucher continued to press the authenticity of the Moulin-Quignon and Menchecourt finds to his death in 1868, with varying degrees of support from the French scientific and archaeological community, but within a decade or so Falconer's condemnation of the finds was completely accepted on both sides of the Channel.

Overall, there are striking similarities between the Moulin-Quignon and the Piltdown frauds - the extraordinary coincidence of not one, but two, finds within the same area and within a comparatively short period of time, the use of sub-fossil skull fragments artificially coloured, the presence of unrelated supporting 'finds' and, above all, the fact that the 'discoveries' provided support at the crucial moment for current geological or anthropological theories and predictions (the antiquity of man in the case of Moulin-Quignon, the simian-like ancestry of man in the case of Piltdown).

Despite its inauspicious precedent in these many respects, the investigators of the Piltdown finds actually carried out _fewer_ tests on their material than did the members of the Moulin-Quignon commission: in particular, tests for artificial colouring and for residual gelatine (which Falconer considered quite crucial in 1863) were not carried out on the Piltdown bones until 1950, nearly forty years after the first specimen was found. What is more, the perpetrator of the Piltdown fraud achieved the success that Falconer and his colleagues denied to whoever was responsible for 'L'affaire Moulin-Quignon'.

Notes

1 K. P. Oakley, "Frameworks for dating fossil man", 2nd edn., London, 1966, p.3; R. Millar, "_The Piltdown Men: a case of archaeological fraud_", London, 1972, esp. chapter VI.

2 See P. J. Boylan, "The Falconer papers, Forres", (a limited edition catalogue with a biographical introduction), Leicestershire Museum Publications, 96 New Walk, Leicester, LE1 6TD, 1977. The original material is now in the Moray D. C. Record Office, Forres, Moray, Scotland. All quotations from items in the collection are cited below in the form 'FPF' together with the catalogue number assigned in the above catalogue.

3 Boylan, op.cit. (2), p.v.

4 G. Prestwich, "Life and letters of Sir Joseph Prestwich", Edinburgh, 1899.

5 C. Murchison, 'Biographical sketch of the author', in H. Falconer, "Palaeontological memoirs", 2 vols., London, 1868, i, xxiii-liii.
6 G. Prestwich, op.cit. (4).
7 J. Evans, "Time and change: the story of Arthur Evans and his forebears", London, 1943, pp.115-18.
8 H. Falconer, letters in The Times, 25 April 1863 and 21 May 1863.
9 Milne-Edwards, 'Note. Sur les résultats fournis par une enquête relative à l'authenticité de la découverte d'une mâchoire humaine et de haches en silex, dans le terrain diluvien de Moulin-Quignon', Compt. Rend. Acad. Sci., 1863, liv, 921-39 (18 May); reprinted in Boucher de Perthes, "Antiquités celtiques et antédiluviennes", 3 vols., Paris, 1864, iii, pp.179-93; H. Falconer, G. Busk & W. B. Carpenter, 'An account of the proceedings of the late conference held in France to inquire into the cirumstances attending the asserted discovery of a human jaw in the gravel at Moulin-Quignon, near Abbeville, including the _procès verbaux_ of the sittings with notes thereon', Natur. Hist. Rev., 1863, xi, 423-62; French version: A.Delesse, 'La mâchoire humaine de Moulin de Quignon', Mém. Soc. Anthropol. France, 1863, ii, 37-68.
10 'On the evidence in the case of the controverted human jaw and flint-implements of Moulin-Quignon', in Falconer, op.cit. (5), ii, 601-25.
11 J. Evans, "The ancient stone implements of Britain', 2nd edn., London, 1897, p.527.
12 Discussed in detail by J. W. Gruber, 'Brixham Cave and the antiquity of man', in M. E. Spiro (ed.), "Context and meaning in anthropology", New York, 1965, pp.373-402.
13 FPF 382 and FPF 383, by A. Faure, Maison speciale de photographies artistiques, Passage du Commerce 13 & 14, Amiens.
14 W. F. Bynum is currently studying the controversy following the publication of Lyell's _Antiquity_ and gave a paper on this to the 1975 INHIGEO Lyell Symposium.
15 Falconer to G. McCall, 8 Feb. [1863], FPF 91.
16 Falconer to G. McCall, 8 Feb. [1863], FPF 91, and 28 April [1863] , FPF 117.
17 L'Abbevillois, 9 April 1863, reprinted in Boucher de Perthes,op. cit. (9), pp.583-4.
18 G. Prestwich, op.cit. (4), p.179.
19 FPF 352.
20 FPF 318.
21 R. Munro, "Archaeology and false antiquities", London, 1905, p.33.
22 C. Murchison in Falconer, op.cit. (5), ii, 602.
23 Falconer, 'The reputed fossil man of Abbeville', The Times, 25 April 1863, p.11.
24 John Evans in Joan Evans, op.cit. (7), pp.116-17.
25 FPF 363F.
26 Falconer, op.cit. (23).

27 FPF 363E.

28 Boucher de Perthes, op.cit. (9), pp.609-13.

29 Ibid. ('Cher docteur, Les bruits qui courent en Angleterre sur la falsification des silex à Abbeville, non-seulement sont contraires à la vérité, mais ils le sont à la raison . . . ') (Acknowledgement is due to L. J. Jordanova and Roger Huss for help with the translations.)

30 Ibid., pp.613-17. ('Pendant ce temps, nos spiristes [sic] de Paris évoquent l'homme fossile d'Abbeville. Dans une séance de spiritisme qui a en lieu jeudi dernier, 23 avril, m'écrit un témoin oculaire et le plus serieusement du monde, l'homme fossile a déclaré que le cataclysme qui a causé sa mort, il y a vingt mille ans, l'avait broyé entre deux pierres avant de déposer sa mâchoire a l'endroit ou je l'ai trouvée, et que son nom était Yoé etc. . . . Le defunt antédiluvien n'a point paru avoir de rancune contre moi, c'est de la générosité, car dans mon livre des Masques, j'ai fait un article contre les esprits et les spiristes . . .')

31 Falconer to J. Evans, 27 April 1863, FPF 363G.

32 FPF 110.

33 29 April 1863, FPF 363H.

34 FPF 363I.

35 Murchison in Falconer, op.cit. (5), ii, 602.

36 FPF 363J.

37 FPF 354A.

38 FPF 363K.

39 FPF 85.

40 Milne-Edwards, op.cit. (9), pp.182-4.

41 FPF 86.

42 Milne-Edwards, op.cit. (9), p.184. ('Mais à ce sujet M. de Quatrefages fit remarquer qu'il pouvait y avoir quelque incertitude relativement au gisement de cette pièce, parce que M. Boucher de Perthes possédait plusieurs dents humaines trouvées dans le même terrain, sur différents points des environs d'Abbeville, et que ce savant, ayant retiré tous ces objets de leurs boîtes respectives pour les montrer en même temps a M. Falconer, craignait de n'avoir pas remis chaque chose à sa place, ce qui pouvait avoir occasionné quelque erreur dans l'application des étiquettes fixées sur ces mêmes boîtes.')

43 Falconer, op.cit. (5), p.615.

44 Milne-Edwards, op.cit. (9), p.187.

45 FPF 87.

46 FPF 88.

47 Ibid.

48 FPF 89.

49 Milne-Edwards, op.cit. (9), p.192. ('Tous les membres de cette réunion d'amis adoptèrent la même opinion. Écartant toute idée de fraude, ils ont reconnu, de la manière la plus franche, qu'il ne leur paraissait plus y avoir aucune raison pour révoquer en

doute l'autenticité de la découverte faite par M. Boucher de Perthes d'une mâchoire humaine dans la partie inférieure du grand dépôt de gravier, d'argile et de cailloux de la carrière de Moulin-Quignon.')

50 Falconer, op.cit. (5), pp.621-2.

('1. La mâchoire en question n'a pas été introduite frauduleusement dans la carrièredu Moulin-Quignon; elle existaint préalablement dans l'endroit ou M. Boucher de Perthes l'a trouvée le 28ième mars denier. Cette conclusion a été adoptée a l'unanimité.

2. Tout tend a faire penser que le dépôt de cette mâchoire a été contemporain de celui des cailloux et autres matériaux qui constituent l'amas argiolo-graveleux, désigné sous le nom de "Couche Noire" laquelle repose immédiatement sur la craie . . . MM. Falconer et Busk . . . réservent leur opinion . . .

3. Les silex taillés, en forme de haches, qui ont été présentés à la réunion comme ayant été trouvés vers la même époque dans les parties inférieures de la carrière du Moulin-Quignon, sont pour la plupart, si non tous, bien authentiques . . . M. Falconer . . . réserve son opinion.

4. Il n'y a aucune raison suffisante pour révoquer en doute la contemporanéité du depot des silex taillés avec celui de la mâchoire trouvée dans la Couche Noire . . . MM. Falconer et Busk . . . désirent réserver leur opinion.')

51 Falconer, op.cit. (5), pp.622-3.

52 Falconer, 'The human jaw of Abbeville', The Times, 21 May 1863, p.11.

53 Ibid.

54 FPF 363L.

55 FPF 363M.

56 FPF 363N.

57 FPF 354B.

58 FPF 363P.

59 See note (9).

60 Boucher de Perthes, op.cit. (9), pp.625-6. ('Ce procès de la mâchoire, comme le nomme le savant et spirituel docteur Carpenter vous a donné bien de l'ennui, et a moi aussi; mais je m'en console, puisque vous m'avez gardé votre amitié, comme je vous ai gardé la mienne . . . Il est pitoyable de voir la passion se mettre dans de telles questions, qui doivent rester dans le domaine pacifique de la science; mais bien des gens, en France comme en Angleterre, en ont fait une affaire de parti: de là tant de colère.')

61 Joan Evans, op.cit., (7), p.118.

62 FPF 363R.

63 FPF 354.

64 FPF 363S.

65 Boucher de Perthes, op.cit. (9).

66 Munro, op.cit. (21), pp.34-5; see also K. P. Oakley, 'The problems of man's antiquity', Bull. Brit. Mus. (Natur. Hist.) Geol. Ser., 1964, ix, 85-155 (111-13).

67 K. P. Oakley, "Frameworks for dating fossil man", London, 1966, p.334.

68 Falconer to G. McCall, 28 April 1863, FPF 117.

69 J. Prestwich, 'On the section at Moulin Quignon, Abbeville, and on the peculiar character of some of the flint implements recently discovered there', Quart. J. Geol. Soc. Lond., 1864, xix, 497-506 (505).

The lost limb: geology and natural history

DAVID ELLISTON ALLEN

'Whatever became of geology?' The question may seem absurd - yet it is a far from unreasonable one for a historian of natural history to ask. For what finally solidified some two hundred years ago into a generally-accepted trinity, an inescapable conjoining of the science of the earth with the sciences of the two kingdoms that make up its modern covering, is manifestly now no more. Geology has continued to flourish; yet in the course of the past century, slowly and unobtrusively, it has slipped out of touch with zoology and botany to the point where communications with them have become all but lost.

Oddly, hardly anything seems to have been written on this massive and far-reaching split. Historians of science have focussed their attention on the internal development of the individual disciplines and have had no eyes for their external contours and their macrocosmic positioning. Yet the way in which disciplines are ordered - or rather seen as ordered - in the wider landscape of learning is surely important in itself. There is a mental cartography at work here which preconditions the extent to which a discipline receives influences from its neighbours. If a science is thought of as closely allied to the one we follow ourselves, we make more of an effort to keep broadly abreast of it, in the hope of borrowing insights, maybe, or in the expectation of coming across findings which fit in fruitfully with what has been found in our own science already. If, on the other hand, a science seems too distantly relevant to make the effort of becoming acquainted worthwhile, we consign it to a mental outer darkness and allow ourselves to forget that it exists. Every discipline has, as it were, its nicely gradated surround of intimates, friends, acquaintances and contacts.

Until the middle of the last century geology, botany and zoology saw one another as intimates - if not indeed closer even than this. A naturalist was anyone who subscribed to the study of the variety and processes of the earth: to be fully deserving of the title, therefore, a person needed to be able to demonstrate familiarity with the subject right across its range. In the eighteenth-century tradition the test of mastery was breadth, not depth; of a philosophy articulated and underpinned over the widest possible span. The aim of learning was roundedness, and the intellectual globe-trotter was still accustomed to travel hopefully. For the purposes of this exploratory essay, I have assumed a degree of homogeneousness in natural history and its component studies that more detailed work may show to be unfounded. To make fruitful comparisons at the gross level of analysis attempted in this paper, some hazardous generalisation forms an unavoidable preliminary. It has been necessary to disregard, for example, the often doubtless divergent attitudes and behaviour of the small minority of professionals as long as this has been a preponderantly amateur field of study - simply because they constitute no more than a small minority, which may accordingly be set aside as exceptional. Similarly, for ease of comparison I have limited my terms of reference to Britain. I am nevertheless conscious that the pattern in other countries will prove in important respects to have been different and that much that is asserted on the strength solely of British experience is going to need substantial qualifying.

'Natural history' is used in this essay in the sense which has prevailed since around 1800 of the scientific study of the surface detail of the earth and of the organisms that inhabit it - in other words, those aspects which are amenable to observation, without requiring recourse to laboratory apparatus. As such, it has been able to remain open to the amateur, indeed in botany and zoology is still predominantly contributed to by amateurs, and so has survived essentially unchanged the general professionalisation which has so profoundly altered the rest of science. For this reason 'natural history' still seems to be both valid and useful as a term.

The conception of natural history as a single entity was greatly aided by the fact that geology's most popular aspect, the collecting of fossils and minerals, was manifestly descended from those very same 'curious' cabinets of the _dilettanti_ as the collecting of specimens of animals and plants. Many fossils were plainly just so many shells, scarcely to be distinguished from those ordinary products of our shores which had once provoked _la conchyliomanie_ and still sustained its attenuated outcome. Others were equally plainly bones - and bones were the province of the zoologist. To a reassuring extent geology thus merged insensibly into these sister facets.

That this undifferentiating outlook bequeathed by the cabinet persisted at least till the very end of the eighteenth century is well shown by the bewildering omnivorousness of the journal kept by Robert Jameson on the occasion of his visit to London in 1793. (1) One moment he is admiring 'a very fine specimen of the Labrador Stone', the next a 'Great crowned Indian Pigeon', then some engravings of plants from Botany Bay, then back again to 'a stalactite from the famous Grotto of Antiparos'. Amongst George Humphrey's 'natural curiosities' his eye lights on 'a beautiful group . . . of Rock Chrystal of a vast size valued at £30.0.0.' and - the very next item - 'a beautiful specimen of the Pennantian Parrot for which he asks £4.4.' All the time Jameson is dashing hither and thither from one great collection to another: one evening to Dr Crichton's fossils, another morning the herbarium of Dr Shaw, the next afternoon Sir Ashton Lever's birds. Almost every corner of creation, it seems, is capable of arousing the enthusiasm of this polymathic nineteen-year-old.

It was not merely the outlook of the collector that preserved this indiscriminateness. The very trend in the study of the earth at that period, with its dominant emphasis on the classifying of minerals, kept the subject in a taxonomic groove and thereby allowed it to continue to resemble the rest of natural history. Indeed, it was doubtless this common legacy of the classificatory approach that was in large part responsible for the study tending in that particular direction in the first place. Certainly it is suggestive that it was the French and the Germans, those much more determinedly system-minded peoples, who took the drive to systematise mineralogy farthest and deepest.

To escape from this sisterly overshadowing, it was necessary for some major new development to occur in the study of the earth which could act as a liberating nonconformity. This in due course arrived in the shape of Huttonian theory. So abruptly unfamiliar was such territory for those rooted in the taxonomic tradition that their disinclination to move into it further opened the way to the emergence at last of a true geology. The later dethronement of the mineralogical approach itself, partly consequent upon the abandoning of the Wernerian system, (2) made this autonomy yet more certain.

Even so, despite having wrenched loose, geology remained just alongside for a further extended period. The relationship had been too prolonged and close to admit of an easy severance. As a result there continued for some while yet some traffic between the three, as botanists tried their hand at geology or geologists sampled zoology.

Unlike before, however, geology now had some dangerous advantages - and not a few who thus wandered over ended up by staying. The very distance the study had achieved from the rest of natural history gave it a slight allure of the exotic. More potently, it also carried the appeal of intellectual forbidden fruit; it was here that the reality of organic evolution gave greatest promise of being uncovered. Above all, it accorded best with the then modish Romanticism. At a period when some acquaintance with science had become essential standard equipment for the upwardly striving, this was at once the most obvious area for newcomers to set about cultivating.

Probably there were homelier causes too. Naturalists are particularly given to being obsessional, and those most deeply obsessed can have found only the one pursuit at a time endurable. Buckland, for example, had started out with the customary schoolboy's addiction to hunting for birds' nests and eggs and might well have continued as a zoologist (as indeed his son later did) had he not chanced upon the more tantalisingly mysterious and (for him) more dashingly eccentric activity of hunting and collecting fossils - and thereupon found it necessary to make a wholesale switch. Jameson's conversion, similarly, was to be a permanent and exclusive one - and regardless of the fact that the chair he occupied was titularly one of a broader Natural History. Lyell, on the other hand, likewise lured across from the botanico-zoological wing, kept a better balance. The son of a well known botanist commemorated in the moss genus Lyellia, he had been primarily an entomologist till won over by Buckland's lectures while at Oxford. In his case the earlier loyalties proved to have taken hold too well for the break to become total; for he was to write frequently on zoological topics in after years and, with lasting fruitfulness, biological considerations never ceased to inform his preponderantly geological thinking.

Returning the compliment, leading botanists like Babington and Henslow faithfully remained Fellows of the Geological Society. This mere deed was not of course demanding and to some extent it must be seen as no more than a token of respect for a sister science that had superior standing. With botany in particular, however, geology had a palpable affinity and it was only to be expected that botanists with broader outlooks should look in this direction in particular. Out of around four hundred and twenty members whom it now proves possible to trace of the ill-documented Botanical Society of London, which flourished for twenty years from 1836, as many as twenty-three belonged to the Geological Society as well and at least another ten are known to have had geological interests. (3) Taking into account the not inconsiderable number whom it has not yet proved possible to identify, perhaps as many as one in every nine of these botanists was sufficiently geologically-inclined to take this to the point of forming a collection of fossils or at least being prepared to meet the relatively heavy cost of a specialist society subscription. These, moreover, were by and large the most active and capable field botanists throughout the country: the mere men-about-town who loomed so large in the metropolitan societies mostly opted instead for the grander and more exclusive Linnean. The extent of the overlap between the Botanical and Geological Societies is thus probably as good a measure as any of the two sciences' contemporary interpenetration. Below this level, in the strictly local and regional societies, separate sections for the different sciences were not as yet the rule and so the proportional allegiances are irrecoverable.

Botanists had a special reason to interest themselves in geology at this period because of the then fashionable concern with explaining plant distributions largely in terms of the underlying rocks. Areas of outstanding appeal to the fossil hunter or the geological mapper also had a marked tendency to be good botanical hunting grounds as well. These two sciences thus had the appearance of being particularly natural bedfellows.

Geology and botany were further akin in representing the non-animal side of natural history: entomology automatically grouped itself with ornithology by virtue of the fact that both of these were concerned with living creatures. Both of the latter, moreover, held out the sporting appeal of a chase. The various major segments of natural history thus mentally ranged themselves along a spectrum, with geology at one end of this, conchology next, botany in the middle, entomology beyond that and ornithology at the other far extreme. The membership of the Botanical Society of London roughly reflects these degrees of affinity in its pattern of secondary affiliations. (4)

As knowledge expanded, however, attempting to ride more than one scientific horse began to become increasingly uncomfortable. To achieve success in doing so, indeed, now became a sign of special genius. Strickland is the outstanding example here, making his mark at much the same period as ornithologist and geologist equally. In the space of a mere fifteen years he not only earned the coveted Readership in Geology at Oxford, but also gave a decisive direction to the study of birds in Britain, converting this 'from a "scientia amabilis" into a serious science', partly by showing the systematists important topics to investigate. (5) Historians of ornithology believe his geological experience may have been partly responsible for his originality.

Yet had Strickland only been spared for a full-length career, it is hard to believe that he would not have settled down as either a geologist or an ornithologist unequivocally. Certainly, in perhaps the best known other contemporary instance, that of Edward Forbes, there is the distinct impression of the intellectual splits. Forbes's first love, throughout his life, was marine biology. Botany was a subsidiary enthusiasm of his schooldays, but this had begun to pall by the time he went to university. Yet, when his father went bankrupt and financial straits forced him (like De la Beche and Robert Bakewell) to turn his recreation into his profession, contrary to all his expectations, the best paid work he could find was a combination of the Curatorship of the Geological Society with the even less remunerative Professorship of Botany at King's College, London. The strain of working in three fields simultaneously, and even more the intense frustration of not being able to give himself wholly to the one that he greatly preferred and to which he looked for his reputation, no doubt played their part in undermining his health and bringing about his death prematurely. (6)

Within a few years versatility such as this had become fatally suspect. Knowledge had expanded to a point where it was no longer possible to make out that there was just the one single discipline with a series of specialised branches. True, it was just at this very same point that a central principle was revealed which enabled these to be grappled together logically far more tightly than before. Yet the unifying effect of evolutionary theory was to be slight in the face of this much deeper functional disintegration. The practical realities which govern the discrimination and demarcation of disciplines are necessarily more compelling than any merely conceptual counter-influence.

In any case, geologists were slow to react to _The origin of species_. As Challinor has pointed out:

> The publication of Darwin's book did not at once cause a rush among palaeontologists to find more evidence as to the manner in which new fossil species appeared in the rocks. One might have thought that attempts in evolutional palaeontology would have been very much to the fore in the eighteen-sixties, but it was not till nearly the end of the century that deliberate studies began to be made in Britain. (7)

Perhaps this is to be seen as just one more indication of the widening estrangement. Geologists had simply become accustomed by then to looking in quite other directions.

As further conclusive evidence of this, geologists and naturalists of other persuasions from around this time cease to exhibit their previous propensity to intermarry. This had been the most concrete testimony of all, that the devotees of natural history of whatever variant constituted a single, indivisible community. What is more, as proof that it was their branch of the subject that had now risen to dominance, it had been the up-and-coming men of geology who had lately been winning the hands of the daughters of natural history's Establishment. De la Beche, for example, had married the daughter of Lewis Weston Dillwyn, co-author with Dawson Turner of the pioneer _Botanist's guide_ of 1805 as well as a prominent conchologist. Strickland, going one better, had become the son-in-law of Sir William Jardine, the wealthy ornithologist and conceiver-cum-sustainer of the long-running Naturalist's library that helped to buttress immediately pre-Victorian zoology. After mid-century no more such alliances are detectable; and the inference must be that geologists had dropped right out of a marriage market that had by now been rendered irrelevant.

Given the vast outpouring of more and more specialised knowledge, it was only to be expected that a greater divergence should have begun to occur, that geology should finally have broken away from the main landmass of natural history and been borne off majestically on its intellectual magma in some other, as yet unpredictable direction. But it was much less clear that this necessarily had to coincide with a severe downward movement in its public following.

Of this decline there can surely be no question. Before mid-century, geologists worked in the full public glare. Even as late as 1858 it was still apparently justifiable to claim: 'There is no branch of Science which attracts so general - it may be said so popular - an interest as Geology.' (8) By the end of the century all but fellow geologists were becoming oblivious of its doings. True, there was a general falling-off in those years of the readiness of the man-in-the-street to try to keep abreast of scientific developments, as discoveries rapidly proliferated and as science as a whole became infinitely more technical and specialised. Yet in geology the waning of interest does seem to have been more than ordinarily widespread and pronounced.

A variety of explanations for this can be adduced. Romanticism had gone off the boil. Terrible monsters had lost their initial shock. Limitless age no longer seemed so impressive. From the 1860s the thrust for origins became man-centred - and that meant a shift to anthropology and in due course to archaeology as well. The turnpikes and the canals had long since been built and by now the majority of the railways too, so there was a decline in conspicuous excavation, with its proneness to generate fossil hunters. The eventual unveiling of evolution had deprived the study of its appealing subversiveness. The big and simple finds had all been made: everything after that was bound to seem undramatic by comparison - and as drama attracts performers so the age of great performers was consequently dead. Most simply of all, geology had been the fashion and, as with all things that fall prey to fashion's spell, the pace had become too hot to last: quite literally the interest had exhausted itself.

Tendencies within the science itself must also have contributed.

Its very maturing intellectually ushered in a general cooling, relieving it of its famed disputatiousness - and thereby ridding it of those hangers-on who had merely come (as Lockhart said of himself) 'to see the fellows fight'. With the healing of those rifts there grew up a greater disciplinary solidarity, with the emphasis on communal endeavour rather than on combat between individual gladiators. The instituting of the Natural Sciences Tripos at Cambridge in 1851 had opened the way to a steadily broadening professionalisation; and young men dependent on their salaries and looking to their careers could not afford the luxury of hitting out regardless in the manner of the gentlemen-scholars. In any case they had far less cause to, for on more and more of the essentials of the study there had now developed a broad consensus. (9) Polemics belonged with the heroics of that now-past age of geology.

In the eyes of the rest of science, too, the subject was experiencing a downgrading. Once it had been the undisputed leader: after mid-century it was doubtfully even the first among equals. For evidence of this, as Kennard (10) has shown, we need look no further than the annual choices for President of the British Association: from 1831 to 1894, eighteen geologists received that honour, from 1894 to 1931 (even allowing for the much shorter period), it was thought worthy of only five. In part this can be explained simply by the greater number of disciplines expecting a share of an unalterably-sized cake, but in part it seems likely to reflect a loss, relative or absolute, in standing. Yet this does not necessarily imply that the subject was in any way deficient at this period intellectually. What it was deficient in, more fatally, was glamour. It was old-established; it was not an experimental study; it was not pursued by members of the new, white-coated priesthood.

The introduction of geology into the curricula of higher and further education only served to widen the severance from botany and zoology. A high proportion of the students who took it were bound for mining or engineering or other bleakly industrial destinations; the world of technology with which they identified was emphatically not the world of natural history. Geology had to fight for its place in the academic sun, and to begin with could thrive only as an ingredient in more general elementary courses. (11) To escape from this treadmill existence, senior teachers inevitably pressed for, and took for themselves, higher-level courses which gave greater opportunities for specialisation. As a result their diet of geology became purer and purer and, often too, more and more esoteric, taking them out of touch with the wider base of their discipline and, more particularly, with the work of their colleagues in the sciences that abutted on it. The splitting of the chair of Natural History at Edinburgh in 1871 (followed, soon after, by a separate laboratory for the zoologists) was symbolic of an alienation occurring more generally. (12) Compounding this, the scholarly action - as with the flare-up of interest in petrology after 1870 - was increasingly in just those areas of the subject that had escaped the attentions of the earlier workers precisely because of their remoteness from, and incongruence with, their essentially field concerns. To this extent there was a parallel with the contemporary opening-up of novel, necessarily laboratory sub-sciences by the burrowers of the New Biology.

Inevitably, too, a growing professionalisation threw up its daunting ring-fence of technical terms and insiders' jargon. Even had the self-taught and non-specialists not been deterred already, this alone was enough to repel them in discomforting numbers. 'Original articles are . . . often too technical to be understood', one of its Presidents was complaining to the Geologists' Association by 1894; (13) 'the advances are made chiefly by specialists many of whom write in cipher, whose ways are dark, and whose language is by no means plain.' It was in the nature of all too much of the subject, moreover, to evolve into a scarcely manageable complexity. By 1886 the total number of fossil species known from Britain alone was estimated at upward of 19,000. (14) No ordinary collector could expect to cope with such overwhelming diversity - and yet if his specimens remained nameless, much of its meaning went out of his pursuit.

Nor did it help that the systematic primary survey of the country, which in all other fields was to present the amateur with a vast and almost inexhaustible challenge, had become in geology the responsibility of a special corps of professionals, who were prosecuting the work with unmatchable expertise and thoroughness. The big expansion of the Geological Survey that occurred in the 'sixties merely served to heighten the impression of a take-over and intensified the mandarin apartness.

It says much for the stolidness of the amateurs who clung on that they were able to make such a success - eventually - of the extra-national body which had been called into being in 1858 specially in response to their needs. (15) So hauntingly similar to the then just-extinct Botanical Society of London as to seem almost its reincarnation, the Geologists' Association, like that predecessor, saw itself from the first as a mutual-aid association for beginners. More deeply, it embodied a current of defiant primitivism: the products of the Working Men's College in Great Ormond Street who banded together to form it felt themselves closer to the ground, truer guardians of the field tradition than the essentially indoor, comparatively pompous Geological Society - which was seen as the home, rather, of 'the illustrious Professors and Masters in the Science' (in the words of the Association's original prospectus). (16)

It was an exact repeat of the split that had taken place in

metropolitan botany twenty-two years earlier. And that it had not occurred till now in geology bespeaks a greater cohesiveness, perhaps attributable to the early lead given by the Geological Society in fostering co-operative work on a nation-wide basis (17) - in marked contrast to the long lethargic Linnean - as well as to the absence of any dispute in geology carried to lengths as bitter as that prolonged ostracism of J. E. Gray by the Linnaeans, (18) and so to the absence of any alternative nucleus of distinction around which a new body could cohere. Contrariwise, the appearance at this point of the very same pattern in geology bespeaks a similarly unbridgeable rift, the arrival at that crux where the discordancy of aims had grown too acute to be contained within a single corporate framework any longer. The out-and-out amateurs accordingly went one way, the out-and-out professionals another - even though, luckily, these were never to become more than polarities: a common allegiance to the field served to bridge the two groups and prevented any more than a partial cleavage.

Indeed, as O'Connor and Meadows (19) have recently reminded us, the development of a separate professional world was uniquely protracted in geology, largely thanks to the vigorous survival of this reverence for investigation out of doors. Geologists, virtually to a man, continued to believe that they were field men or they were nothing. It never occurred to them to copy the new wave of botanists and zoologists and seek to accentuate their newness by dismissing all such work as the stigma of a discredited past.

It is almost impossible for geologists, with their serene inheritance of the field tradition, to conceive of the hostility, even downright virulence, with which this was assailed and rejected by the upholders of the new laboratory creeds of the botany and zoology of the late Victorian period. (20) Admittedly, the teaching of the rudiments of systematics had descended to a peculiar nadir of aridity at the hands of their professorial predecessors. Admittedly, old-style scholars like Babington at Cambridge did cling to their chairs and the coveted departmental funds with an exasperating stubbornness. Yet it is hard to accept that these frustrations warranted such a prolonged display of iconoclastic fury. And it was iconoclasm in very truth: much that was wilfully stamped upon was in its way of great value and for all too many years had no hope of being replaced.

Starved of professional inspiration, cut off from the flow of potentially invigorating new knowledge, field botany and zoology consequently stagnated and remained almost wholly amateur studies. As a result, the image of mustiness which their detractors hastened to fasten on them came for a time all too uncomfortably close to the reality. Sapped, moreover, by sentimentalism, which the scholarly element had become too enfeebled to keep suppressed, they acquired a reputation that was doubly unfortunate.

All of this geology escaped. And the very fact that it did escape it is conclusive testimony in itself that the break with the rest of natural history had by then become more or less complete.

Ever since, geologists on the one hand and botanists and zoologists on the other have stared at one another across a chasm. The Geologists' Association and the Botanical Society of the British Isles, two of the largest bodies in their respective fields, have evolved along quite remarkably similar lines into structurally almost identical organizations - yet have only ever been in contact once, and then quite informally. Of the many general natural history societies that today exist in this country scarcely any even attempt to embrace geology, let alone boast of a section specially devoted to it. Even in the avidly proselytising world of conservation there is doubt whether the bounds of natural history can now credibly be pushed out as far as this. 'Some Trusts', runs a statement in a recent newsletter of the Society for the Promotion of Nature Conservation, (21) 'might even feel that geology is not within their brief.' However, it goes on, 'this would be unfortunate'. Only in the inherently antique world of museums, most of them local ones, has the old unity been granted preservation - and mummification might be a truer, if unkinder word for this.

Yet, viewed from a different angle, the separation appears unreal, a mere artefact of just one conceptual scheme in particular. The landscape which has been depicted in this paper is a purely cognitive one: the positioning of the disciplines is their positioning merely from the standpoint of scientific knowledge. If, instead, they are viewed from the standpoint of social function, we are confronted with a very different picture. Class geology as a field study, group it with the now-reinvigorated field aspects of zoology and botany - with ecology and ethology, with population genetics and experimental taxonomy - and what do we have but a contemporary cluster of ways of directly investigating the environment which is no more and no less than an updated natural history. In this sense of a series of intrinsically cognate activities the old unity remains unchanged. Despite their divergence intellectually the three disciplines still share a core of attitudes and functions, which has the effect of keeping them together. From the sensing of this truth, even though it eludes rational articulation, spring those continuing, puzzled moves to reassert the threeness of natural history.

It has been traditional up to now for historians of science to subscribe exclusively to the cognitive scheme of things. But as the viewpoint of the social historian impinges, a shift to alternative forms of categorisation can be expected. To define disciplines in terms of how they are organised and of the modes in which those who belong to them carry out their work, instead of in terms of their particular siting within the landscape of knowledge, may even be more helpful for those who have to take the decisions for their institutional nurture and support. Sciences may march together in the pattern of their activities while standing far removed in the paradigms that bound them intellectually. To regard geology, therefore, as an integral part of a wider 'field science', over and above its existence as a discipline in itself, would seem to be as defensible as it may prove to be conceptually beneficial. For by looking at it as part of this other, wider whole, we may hope to arrive at a deeper understanding of it - no less than of natural history too.

Notes

1 J. M. Sweet, 'Robert Jameson in London, 1793', Ann. Sci., 1963, xix, 81-116.

2 R. Porter, "The making of geology", Cambridge, 1977, pp.170-6.

3 Unpublished work by the writer still in progress. This appears to be the only metropolitan society of this period which has been the subject of an intensive membership analysis.

4 In so far as data on this point are now recoverable, at least 34 members of the Botanical Society were also entomologists (as against 43 with geological interests) but only 14 were also ornithologists. Anomalously, no more than 14 are revealed as conchologists, though a further 5 took in marine zoology in some other form.

5 E. Stresemann, "Ornithology: from Aristotle to the present" (tr. H. J. & C. Epstein, ed. G. W. Cottrell), Cambridge, Mass. & London, 1975, p.225.

6 G. Wilson & A. Geikie, "Memoir of Edward Forbes", Edinburgh, 1861, passim.

7 J. Challinor, "The history of British geology: a bibliographical study", Newton Abbot, 1971, pp.139-40.

8 "The Geologists' Association 1858-1958", ed. G.S. Sweeting, Colchester, 1958, p.2.

9 R. Porter, 'Geology in the Cambridge Natural Sciences Tripos 1851-1914', unpublished paper, p.8.

10 A. S. Kennard, 'Fifty and one years of the Geologists' Association', Proc. Geol. Ass., 1948, lviii, 271-93.

11 R. Porter, 'The development of the teaching of geology in Britain 1660-1914', unpublished paper, p.10.

12 J. B. Morrell, 'The patronage of mid-Victorian science in the University of Edinburgh', Sci. Stud., 1973, iii, 353-88.

13 H. B. Woodward, 'Geology in the field and in the study', Proc. Geol. Ass., 1894, xiii, 247-73.

14 F. W. Rudler, 'Fifty years' progress in British geology', Proc.

Geol. Ass., 1887, x, 234-72.

15 T. R. Jones, 'The Geologists' Association: its origin and progress', Proc. Geol. Ass., 1883, vii, 1-57.

16 See note (8).

17 M. J. S. Rudwick, 'The foundation of the Geological Society of London: its scheme for co-operative research and its struggle for independence', Brit. J. Hist. Sci., 1963, i, 324-55.

18 A.E. Gunther, "A century of zoology at the British Museum through the lives of two keepers, 1815-1914", London, 1975, p.44; J. E. Gray, 'Sowerby's "English botany"', J. Botany, 1872, i, 374-5.

19 J. G. O'Connor & A. J. Meadows, 'Specialization and professionalization in British geology', Soc. Stud. Sci., 1976, vi, 77-89.

20 F. O. Bower, "Sixty years of botany in Britain (1875-1935): impressions of an eye-witness", London, 1938, provides the best insight into this. The subject is treated at greater length in chapter IX of my "The naturalist in Britain: a social history", London, 1976, pp.180-4.

21 P. Toghill, 'Geological conservation', Conservation Rev., 1976, no.13, p.2.

The social history of geology

Geological communication in the Bath area in the last half of the eighteenth century

HUGH TORRENS

> I am as delirious as ever, still preferring a coal-pit or stone quarry to the Bath Assembly, or a Court Ball. (1)
> [E. M. da Costa to Ralph Schomberg of Bath, 1761.]

Introduction

Records of geological activity in the Bath area in the last half of the eighteenth century are certainly relatively sparse. (2) There are two main reasons for this. Firstly, much of the activity would have been devoted to collecting geological specimens and especially fossils; such collections, however much treasured by their makers, have long since been dispersed and records of this activity are likely only in references in letters. Secondly, even the records of the organised geological activities of the first two Bath Philosophical Societies, which both flourished in the late eighteenth century, have similarly been dispersed and scattered to the winds. Despite such massive erosion of the basic source material, a study of the Bath area still seems well worth attempting for two reasons. First, the special place Bath occupied in the last half of the eighteenth century 'as the most fashionable place for idling and the most famous watering place in Britain', (3) where one would imagine the collecting of facts and specimens of geological significance would naturally take its place along with the other multifarious activities of the numerous visitors. Second, the special place Bath has occupied since 1827 as the so-called 'cradle of English geology' from the work of William Smith (1769-1839) in this area from 1791 to about 1805. (4)

The special nature of Bath society in the eighteenth century

Despite the massive amount of secondary literature Bath has engendered, there is still 'no professional economic or social history of Bath' in the eighteenth century. (5) There is, however, massive documentation of the popularity of Bath as a resort for medical aid or leisure activity. The most eloquent surviving testimonies to the former are the monumental inscriptions to the many visitors who also died in Bath, which to this day line the walls of Bath Abbey. (6)

These walls adorned with monument and bust
Show how Bath Waters serve to lay the dust
Henry Harington (1727-1816)

The leisure life of Bath was described thus at its zenith in about 1791: 'No place in Europe, in a full season, affords so brilliant a circle of polite company as Bath. The young, the old, the grave, the gay, the infirm and the healthy all resort to this vortex of amusement. Ceremony, beyond the usual rules of politeness, is totally exploded.' (7) Barbeau (8) has described the range of social and literary activities available in Bath; music, (9) the theatre, (10) and bookselling (11) all thrived. The surviving editions of guides, directories, and newspapers produced for the visitors to Bath give a further indication of the size of the leisure industry, and there is a considerable if disseminated literature available on some of its different aspects.

The scientific activity has however been largely ignored. Barbeau stated of Bath inhabitants in the eighteenth century that 'there was little desire for literary or scientific knowledge', (12) but this is certainly not true of a nucleus of Bath residents and visitors who took an active interest in a wide range of aspects of literature and natural philosophy. (13)

Many people from Warner (14) onwards have also claimed that Bath lacked a solid trading and manufacturing base (15) at this time and that industry - that vital catalyst - was absent from Bath. While this may be true of heavy industry, it is certainly not true if, with J. H. Plumb, (16) one considers the tourist trade as an industry. As R. S. Neale has shown 'it was a city where a multitude of skills were in high demand to build, maintain, furnish, feed, clothe and entertain its wealthy residents and visitors. The city was a paradise for the consumer industries', (17) and Warner's claim that Bath in 1801 had no manufactures and little trade can be easily discounted for this single reason. The activities of one closely knit group of tradesmen who settled in Bath in this period from many different places has been briefly documented; (18) they included ironmongers, wine merchants, corn factors, brewers, coachbuilders, and pleasure garden proprietors - all representative of the very important service industries so well developed at this time in Bath.

There were indeed two major service industries which were both vital to the expansion of Bath and the level of geological activity in the city. These were the building industry (19) and the stone quarrying industry, (20) both of which would have had profound geological significance and afforded 'much speculation for the naturalist and virtuoso' as Stebbing Shaw noted (21) of the fossils and materials thrown out of the foundations for the Lansdown Crescent built from 1789 to 1793.

Geological activity

Bath enjoyed a major boom in the mid-eighteenth century when new building on a large and beautifully planned scale produced one of the most fashionable resorts in Europe. Bath had been popular before this on account of the mineral waters and the hot and cold baths. John Woodward (1665-1728) had been a medical visitor there in August 1722 (22) and was a friend or at least acquaintance (23) of the Bath physician Thomas Guidott (1638-1705) who practised here for nearly forty years (24) and wrote much on the Bath waters, and others nearby. Another important scientific figure with early Bath connections is John Theophilus Desaguliers (1683-1744), (25) early itinerant science lecturer and curator of experiments to the Royal Society from 1716 to 1743. He was a frequent lecturer at Bath and resided here at intervals, being a constant visitor to the Royal Cumberland Lodge of Masons established at Bath in 1733. (26) In the following year, he published a detailed description of the Bath quarry tramroad built in 1731 by Ralph Allen (1694-1764) to run from his Combe Down quarries down to the River Avon. (27) This expansion in quarrying the local Bath Stone was to meet the demand created by the activities of John Wood the elder. Desaguliers provided the first detailed description of any railway system in English, (28) and he also gave details of the cranes then in use in the quarries.

All this activity in expanding the town of Bath would have encouraged the study of the embryonic science of 'geology', in that it would at least have greatly facilitated the direct observation of rocks and their obvious stratification in the many excavations needed, as well as the collecting of fossils. Alexander Pope (1688-1744) was one person who was considerably encouraged in his collecting; he stayed some months in 1739 with Ralph Allen at Widcombe near Bath. Pope had some years before, in 1725, finished his grotto at his Twickenham house, which fancifully used shells, flints, iron ore, mirrors, or whatever came to hand to create a place of surprise. After his visit to Bath, Pope resolved to improve it and add some Bath curiosities provided by Allen such as Bristol diamonds, alabaster, spars, and snakestones (ammonites), which were all duly incorporated. (29)

More scientific study of the rocks of the area came from John Wood father and son. John Wood the elder (1704-1754) was the architect and builder who changed Bath 'from a mean looking town to the most beautiful in England' and it is no surprise in view of his vocation to find him devoting chapters on the 'Situation of Bath; of its vales and of its hills' and on the 'Soil of Bath and the fossils peculiar to it' to his _Description of Bath_, which we can take as a summary of Bath geological knowledge and interest at the mid-point of the century. (30) It contains geographical and geomorphological material of no particular note and details of the Bath springs, which were the major attraction to so many visitors. Wood has been criticised for his absurd credulity as to past history, but he was not being more credulous than many others of his time in writing 'in the formation of the Hills that surround the hot springs, Nature seems to have had a spiral Motion so as to form a kind of Volute', an idea found in the writings of many others before and after. Wood connected with this 'the spiral figures [ammonites] which I shall hereafter show to be peculiar to the soil of Bath and perhaps no where else to be met with in such great abundance and in such infinite variety'. (31) The chapter on the soil and fossils is of more interest as it gives rudimentary stratigraphic information and quotes and criticises the late John Strachey (1671-1743) on the local coal mines. Wood observes that the soil near Bath abounded with fossils 'mostly with such as are of a Spiral Figure and such as our Naturalists believe to have been formed in Nautili Shells', and also 'multitudes of conical stones with elliptical bases found in almost all the stratas of clay and marl within Bath' (i.e. belemnites). (32) He notes also the beds of gravel abounding with different round fossils (i.e. flint echinoids) and concludes 'many other little miracles of nature abound in the soil of Bath to excite a Man's Curiosity to examine into them; and an Age may be spent in a Pursuit of this kind so abundant are the Fossils wherever the Ground is penetrated . . . ' (33)

Wood mentions no contemporary collectors other than himself, but another writer, Charles Lucas (1713-1771), in his 1756 _Essay on waters_, paid tribute to one of the most active of the Bathonian geological virtuosi of this time, Thomas Haviland. (34) He was an apothecary born circa 1706 probably in Bath, and first heard of in 1740 as a subscriber to Thomas Short's _Essay on mineral waters_. By 1753, he had started an extensive correspondence with E. M. da Costa which continued until 1762; (35) he was also a friend and correspondent of traveller and naturalist Thomas Pennant (1726-1798) to whom he gave fossil bivalves 'found in a stoney bank near

Bath', (36) and of naturalist John Ellis (c.1705-1776) with whom he corresponded on botany as well as geology in 1755. (37) In 1756, Lucas described Haviland as an excellent apothecary and an accurate botanist with an extensive collection of local Bath fossils as well as others from further afield. His fossil collections seem to have been his main interest, to judge by the Gainsborough portrait of him painted about 1762 which shows two Lower Lias molluscs prominently on his bookshelf and a 1761 edition of Lewis's _Materia medica_ in his hand. (38) There is sufficient evidence to show that Haviland, who died in 1770, was both an active and diligent collector and a classifier of fossils. (39)

Almost contemporary with Haviland in Bath was the physician Ralph Schomberg (1714-1792) who settled in Bath in the 1750s. He was another of da Costa's correspondents and also a friend of Haviland. (40) He had previously been in medical practice in Yarmouth, being resident there in 1752 when he was elected a Fellow of the Society of Antiquaries in London. Da Costa specifically sought his help from Bath in augmenting his collections as Schomberg 'was in a place - the quarries of which abound in figured fossils'. (41)

It was also Schomberg who recommended John Walcott the elder (died 1776) to Fellowship of the Society of Antiquaries in 1766 - his certificate stating he was well versed in history, the _belles lettres_ and antiquity. (42) He came from Ireland, owning estates at Croagh near Limerick, a city of which he was made a Freeman in 1750. (43) In 1753 he married a Cork merchant's daughter and his second son named Edmund was born there in 1756. (44) Sadly, the relevant parish records have been destroyed so we are unable to confirm the place and exact date of his eldest son John's birth - who has been claimed as the first Bath geologist. John Walcott senior brought his family to Bath some time between 1756 and 1766 while living off the rents from his Irish estates. Under these conditions his eldest son John fostered an eager interest in all branches of natural history. In a deed of 1759 he is described as about four years old which would suggest he was born in 1755 (45) and agree with his next youngest brother's known birth in 1756. Their father died in 1776 (46) and left evidence of wide scholarship in his library sale catalogue (47) which amounted to 1,675 lots and contained most of the standard works on natural history and geology of the time. His nonconformist attitude was reflected in his burial entry in the registers of Weston near Bath. (48) This trait reappeared strongly in the eldest son, John Walcott junior.

It must have been largely due to the father's influence that John junior took up the study of natural history. He published his first work in 1778-9 in fourteen monthly parts; this was a never completed British Flora which had the laudable intention of providing accurate engravings of the common British plants at a price within the pocket of all but the poorest. (49) Despite encouragement from Erasmus Darwin, it was discontinued in 1779, presumably for lack of subscribers. (50) John Walcott's next book appeared in July or August 1779 at two shillings and six pence, when he was still only twenty-four years old. (51) It was entitled _Descriptions and figures of petrifications found in the quarries, gravel-pits, etc. near Bath_, and deserves careful attention as it was acknowledged and used by Bath's most famous later geologist, William Smith. (52) It is prefixed by a quotation from Benjamin Stillingfleet's _Miscellaneous tracts relating to natural history_, a work which had been in his father's library: (53)

> Nor are those innumerable petrifactions, so various in species, and structure, to be looked upon as vain curiosities. We find in our mountains, and even in the middle of stones, as it were embalmed, animals, shells, corals, which are not to be found alive in any part of Europe. These alone, were there no other reason, might put us upon looking back into antiquity, and considering the primitive form of the earth, its increase, and metamorphosis. (54)

This book contains descriptions and excellent figures drawn by Walcott himself of the many fossils 'found lodged in stone in almost every part of the environs of Bath'. (55) It ascribes all these to a former universal deluge and displays an impressive grasp of the literature. It classifies these fossils into several groups, such as (i) internal moulds of bivalves of which Walcott observed, 'shells are never filled with stone different from that in which they are lodged', or (ii) ammonites, which 'as far as I have observed the flats of these [_Coroniceras_] lay parallel and conformable to the surface of the stratum in which they are enclosed'. (56)

Localities are, however, rarely given and then never with the accuracy we would hope for; no stratigraphic information is given. He refers merely to freestone [equivalent to the Great Oolite limestone] or limestone [equivalent to the Lias] or simply gives localities as on 'the Ploughed Fields'. He is at his best when seeking affinities with living shells, some of which he figures side by side with the fossils. (57) He also confesses himself defeated by the affinities of some fossils now known to be extinct.

Although in no way breaking new ground, it was and is a remarkable achievement for a twenty-four year old, especially for the accuracy of the drawings, which allows the greater proportion of the forms figured to be identified to specific level today. This accuracy of observation and illustration is nowhere better shown than in his _Spiriferina Walcotti_ (a Lias brachiopod named in his honour by James Sowerby in 1822) (58) in which the calcified internal spiralia are beautifully shown.

Although it cannot be proven, Walcott's personal contacts with other scientists outside Bath seem to have been very limited. References to him in manuscript collections are very few and uninformative and he is remarkably badly known for a man who produced four natural history books. One of the most likely influences on his early work in botany and geology was Edward Jacob's (1710?-1788) botanical book published in 1777. Jacob was a surgeon from Faversham, Kent, (59) who began collecting London clay fossils in the 1740s and published on them in the _Philosophical Transactions_ in 1754. Like John Walcott senior, he was a Fellow of the Society of Antiquaries. (60) The main influence on Walcott junior seems to have been Jacob's _Plantae Favershamienses_ (1777) rather than any personal contact; this last book shows such a similar style of presentation to Walcott's more impressive productions, even down to an identical Biblical quotation on the two botanical title pages. _Plantae Favershamienses_ has a short 'Appendix exhibiting a short view of the fossil bodies of Sheppey' and this too seems likely to have influenced Walcott's book on Bath petrifications two years later. (61)

In November 1779, a few months after Walcott's book had been published, Caleb Hillier Parry (1755-1822) settled as a physician in Bath where he soon built up one of the most prosperous practices. (62) This, sadly, is the reason no more is heard of Parry's intended work on the fossils of Gloucestershire. He issued printed _Proposals for a history of the fossils of Gloucestershire_ in 1781 - no copy of which appears to survive. But both John Britton, (63) the topographer and later friend of William Smith, and Parry's son, Charles Henry Parry (64) referred to it, mentioning work Parry had devoted to it before first his extensive practice and then other scientific and agricultural interests caused him to set it aside. It was intended to include all that was known on the subject of organic remains and the discovery of the original manuscripts which were in existence in 1830 would be of major importance. (65)

Parry built up extensive collections of fossils which were later augmented by his purchase of those of William Cunnington (1754-1810) of Heytesbury, Wiltshire in 1810. (66) We can only assume that Parry's birthplace (Cirencester), his scientific training at Warrington Academy and Edinburgh University, and the publication of Walcott's book so soon before his arrival in Bath, all aroused his interest in fossils. Parry certainly possessed a copy of Walcott's book which contains the later signatures of his son and S. P. Pratt, two further generations of Bath geologists. (67) In 1820, his son Charles Henry Parry (1799-1860) recorded in his manuscript autobiography a visit to Toghill near Bath, 'once famous for its quarries and specimens - recorded by my Father in his MSS. notes to Walcott's Petrifications', (68) confirming that Walcott's book was indeed a major influence on Parry's work in geology.

In December 1779, a letter from Thomas Curtis (c.1739-1784) to

Edmund Rack (1735-1787), both then living in Bath, led to the formation late in that year of the first Bath Philosophical Society. (69) John Walcott was a founder member while Caleb Parry had been elected a member by March 1783. (70) This Society has been strangely ignored, but it certainly predates that at Manchester founded 1781 and often wrongly claimed as the oldest of the provincial philosophical societies. (71) Little is known about Thomas Curtis, a gentleman of means and well applied leisure whose idea the Society was; but it bore instant though short lived fruit under the enthusiastic secretaryship of the Quaker Edmund Rack. Rack had moved to Bath in 1775 better to enjoy the literary life there, and he was soon energetically channelling the indigenous and immigrant talent of the place into an Agricultural Society which celebrated its bicentenary in 1977. (72) The Philosophical Society (73) was a private society of never more than twenty-five ordinary members. (74) Among these for a short while were men of the calibre of Joseph Priestley (1733-1804) (75) and William Herschel (1738-1822), and it is the latter's membership until 1782 which provides much of the information we have about the Society. The thirty-two papers Herschel submitted to the Society on physics, metaphysics, and astronomy have largely been published (76) and give the quite false impression that the Society was not at all concerned in studies of natural history. New material shows this to be wrong, as it equally shows that technological discussion at the Society was minimal, unlike the contemporary Lunar Society which Priestley later joined on moving from Calne in Wiltshire to Birmingham in 1780. (77)

Members with an active interest in natural history included William Watson junior (1744-1824) MD Cambridge 1771, FRS 1767, who was related by marriage to Samuel Galton junior (1753-1832) of the Lunar Society and who supported Galton's certificate for Fellowship of the Royal Society in 1785. Watson published on zoology (78) and was keenly interested in botany and geology. (79) Caleb Hillier Parry (1755-1822) MD Edinburgh 1778, who has already been mentioned, as has John Walcott (1755?-1831), were also members, along with John Walcott's friend Matthew Martin (1748-1838) who published several works on both botany and entomology and who contributed natural history papers to the Society. (80) Martin was a friend and correspondent of William Withering, one of the major naturalists of the Lunar Society. (81) John Coakley Lettsom (1744-1815) MD Leyden 1769, FRS 1773, and Matthew Dobson (died 1784) MD Edinburgh 1756, FRS 1778, (82) were two members of the Society who were active in geology. A full study of the membership and activities of this and later Bath Philosophical Societies is in progress by the author, but preliminary evidence already shows that fossils and their origins were discussed at Society meetings following an initial paper by Thomas Parsons (1744-1813), who was a Baptist minister in Bath and also, like his better known father Robert (1718-1790), a stone mason and carver. (83) This latter activity, which Thomas continued up to 1791 at least, would explain a knowledge of, and interest in fossils. (84)

Edmund Rack also seems to have spent a good deal of what little free time was left over from acting as Secretary to both the Philosophical and Agricultural Societies thinking about, discussing and collecting in natural history and geology. In May 1779, on a journey to Sidmouth in Devon collecting subscriptions for the Agricultural Society in Bath, he collected some living corals at the coast. In August 1779, he wrote about these to the Gentleman's Magazine, (85) and on 13 January 1780 they formed the subject of the second paper submitted to the Bath Society, while they were further discussed in William Herschel's first papers to the Society. (86) Rack submitted his mature reflections on these animals to the Royal Society in 1782. (87)

After the subject of fossils had first been presented by Thomas Parsons, as the fifth paper to the Bath Society on 21 January 1780, Rack himself seems to have devoted much attention to the subject. Perhaps the best demonstration of this is the entries in the Bath Agricultural Society Accounts from 1780 onwards which include the following entries: (88)

1780 June 28 Expences of Journey into Dorsetshire to collect Subscriptions to the Agricultural Society and Natural Curiosities. 10 days	£4-15-0
1780 Sept. 10 To Fossils	1-3
1780 Sept. 14 A Box for Fossils	1-6
1780 Sept. 24 Carriage of a Box of Fossils from Norfolk	2-2
1780 Oct. 24 Carriage of Sea Shells and Fossils - Weymouth	2-6
1780 Nov. 10 Carriage of Basket Fossils from Charmouth	2-9
1780 Nov. 16 Carriage of Basket from Charmouth	1-7
1780 Dec. 9 Carriage of Fossils and Petrefactions from Portland	4-4

After this entries become less specific and we find instead:

1781 May 26 Journey into Dorsetshire collecting subscriptions	£ 1-9-9
carriage of a Box	2-6

This journey is again closely followed by the carriage of a box which we may assume again contained minerals (fossils to Rack) or fossils (petrifactions to Rack) as before, and that subscriptions were not the only thing Rack was collecting in Dorset!

In 1781, Rack and the Rev. John Collinson (1757-1793) issued the first proposals for their intended History of the County of Somerset. (89) Further proposals seeking subscriptions were issued in 1784 and 1785. (90) Rack's part in this work was to survey the topography and natural history of each Somerset parish, which he did assiduously from 1781 to 1786. Rack died before the work was published in three volumes in 1791 by Collinson and 'so has almost lost the credit due for his work'. (91) The recent discovery of Rack's original manuscript for this work (92) shows what a great deal of work he had accomplished and allows his major share of this work to be correctly assessed. His travels over Somerset for this project gave him opportunities to travel widely in the south-western counties and to meet fellow fossilologists. One such was the Rev. John Wickham (1730-1783) of Horsington in South Somerset. He had graduated BA at Oxford University in 1749 where he had attended James Bradley's lectures on natural philosophy (93) and made the acquaintance of William Huddesford (1732-1772), curator of the Ashmolean Museum, (94) to whom he sent fossils from the Dorset coast in 1760. Wickham was also in contact with the naturalist Gustave Brander (1720-1787) (as was John Walcott), (95) and William Curtis 1746-1799, the Quaker botanist. (96) Wickham was a most enthusiastic collector of fossils, as both Smart Lethieullier (1701-1760), who in 1760 described him as 'as fairly catch'd in the fossil trap as any one I have ever met with', (97) and his monumental inscription in Horsington church testify. Wickham seems to have been the rediscoverer of the celebrated Liassic ammonite marble of Marston Magna, Somerset, specimens of which he presented in 1782 to the Woodwardian Museum in Cambridge. Information about his discovery of this marble was transmitted in 1784 by his friend Rack to both John Hunter (98) and to the Society for Promoting Natural History, in London, founded in 1782. (99)

Other members of the Bath Philosophical Society with an interest in fossils included an important member of that particularly eighteenth-century phenomenon, the itinerant lecturer in science, called John Arden (1720-1791). (100) After an active career lecturing and teaching throughout England, he settled in Bath in 1777 for about six years. At this time he was already the friend of many well known scientists such as Priestley (a fellow Philosophical Society associate) (101) and Josiah Wedgwood. (102) He had previously lectured in Bath at least twice in 1769 and 1770, (103) and he must have found audiences and life here congenial for we find him as a founder member of the Agricultural Society, soon after his arrival in Bath, to which he soon offered the use of his 'very complete Apparatus of Philosophical Instruments', (104) and two years later in 1779 of the

Philosophical Society. In 1777 we also find his eldest son, James (1752-1842), corresponding a little acrimoniously with E. M. da Costa about the purchase of a 'fossil' collection of over 700 specimens both 'native and extraneous', which arrived from da Costa in August. (105) It seems at least possible that John Arden may have used these in his Bath lecture courses which Edmund Rack attended, although no mention of such lectures appears in his printed synopses. He also may have delivered special lectures on fossils and minerals as da Costa was doing at the same time in London. (106)

In 1782 the Society for Promoting Natural History was founded in London. 'Mr John Hamlyn - Miniature painter at Bath' was admitted an honorary member at the second meeting. He was merely interested in geological specimens as curiosities, to judge by the flint from his collection figured in 1781. (107) Other members of this society with strong Bath connections were Edmund Rack (honorary member 1783), John Walsh of London (ordinary member 1785), John Coakley Lettsom of London (ordinary member 1786); both Walsh and Lettsom were already members of the Bath Philosophical Society. Another honorary member (1784) was James Stephens of Camerton near Bath, who was one of William Smith's earliest patrons and one of the chief instigators of the Somerset Coal Canal which was of such significance in Smith's career. Rack contributed at least two papers to this Society on fossils and their origins, which must also have been read to the Bath Society. The London society was 'much interested in exploring fossil bodies'. (108) Among the bodies they explored were the jaw of an unknown animal with many teeth from the Dorset coast (perhaps an Ichthyosaur) exhibited by Hamlyn and fossil plants sent by Stephens from Camerton, which had been found in sinking one of his collieries, later to be served by the Coal Canal. Dr Henry Menish (died 1810) of Chelmsford, who was elected an honorary member of the London society in 1785 also appears in contact with Rack before this time; the Bath Agricultural Society accounts record, in Rack's hand, their payment of the postage on a 'Packet from Dr Menish' on 24 July 1784, no doubt again containing fossils for the use of Rack and the members of the Bath Philosophical Society. (109)

For the period from 1777, with the foundation of the Bath Agricultural Society, through 1779, with the foundation of the Philosophical Society, Rack appears as the focal point of scientific activity in Bath, and especially in its dissemination outside the area, until his death in 1787. Apart from the Society for Promoting Natural History, he was in contact with other Agricultural Societies, as well as the Society of Arts, the Royal Society in London, and the Literary and Philosophical Society founded in 1781 in Manchester. After 1779 the German, Georg Lichtenberg (1742-1799), would surely never have written to Herschel himself in early 1783 of his earlier visit to Bath, 'Good Heavens! had I but known, when I spent some days in Bath in October 1775 that such a man [i.e. Herschel] was living there'. (110)

Not only did the Bath Philosophical Society bring Herschel forward, it also effectively improved scientific communication to and from Bath between the inhabitants and numerous visitors to Bath, (111) through its indefatigable secretary Edmund Rack. With his illness in 1786 and death in early 1787 this channel of communication ceased and the Society was dissolved. (112)

Several areas of geological communication have not yet been touched on. We have little record of the role of museums in the area. Certainly by this time travelling displays of natural curiosities would have been visiting Bath, like Rev. Robert Ferryman's primarily zoological collection in April 1789, (113) but of these we have till now little evidence. A museum was set up in Bristol in 1784 at the Baptist Academy there (114) and certainly contained fossils and other geological material by 1799, but the first reference to such a collection on public display in Bath is not found until 1809 when the largely Natural History Museum at 21 Union Street was open. (115) Similarly little is known about the sale of geological specimens in Bath at this time. Rack's Journal mentions private collections of fossils and minerals apparently for sale in Bath and of these details can sometimes be found in the advertisement columns of the local press (see Illustration 9). (116) The first dealer as such to be located is James Lintern, listed as 'Music and Petrefaction Warehouse, Abbey Yard' in the 1787 Bath Directory. He died in February 1817 and was, with his brother Walter (who died in 1806) a well known publisher of music and a musical instrument maker. (117)

Casual visitors to Bath, of which there were multitudes at this time, may also have widened the circles of geological communication, both in the area and outside. Details of visitors are sparse and we know no more of James Hutton's (1726-1797) visit here than that it took place in September 1774. (118) Jean André de Luc (1727-1817) was a frequent visitor from 1773, (119) his first year in England, and in 1785 he married a Bath widow at St James church in Bath. (120) In 1787 he too noted the regular stratification and abundant fossils, (121) temporarily exposed in the feverish building activity on the slopes of Lansdown north of Bath, which were commented on by several other writers. (122) Others resident in Bath for short periods could be mentioned, such as David Erskine, eleventh Earl of Buchan (1742-1829), who lived in Bath until the death of his father there in 1767, and who was a friend of Alexander Catcott (1725-1779) of Bristol and the dedicatee of the second edition of Catcott's Treatise on the deluge (1768). He sent Catcott word in 1766 of the discovery of a forty foot 'Young Whale' in the Lias of Weston, near Bath. (123) Another short-stay resident of Bath was the physician, Dr John Berkenhout (1730-1791) MD Leyden 1765, (124) who lived and practised here between 1781 and 1782, meeting William Herschel with whom he corresponded in 1788. (125) Berkenhout had published an influential

BATH, MAY 1, 1783.

TO be SOLD by AUCTION,

By WILLIAM CROSS,

On Wednesday the 14th instant, and following days,

At the Exhibition-Room in Bond-street,

All the PICTURES, PRINTS, DRAWINGS, MEDALS, COINS, FOSSILS, and SHELLS, the property of

Mr. HAMLYN, (who is going Abroad.)

Among them are the Works of the following Ancient and Modern Masters, viz.

Della Bella	Vandeist	Guercino
Polydore	Cuyp	P. Veronese
Antonio	Teniers	Hemskirk
Callott	Grimm	Titian
Poussin	Gainsborough	Carlo Maratti
Brueghel	West	Vandyke
Spagnoletto	Worlidge	Carracci
Parmegiano	Raphael	L. Da Vinci, &c.

A capital picture of Moses and the Children of Israel at the foot of Mount Horeb, by Parmegiano; beautiful Drawings elegantly framed and glazed.

The Medals, Coins, &c. are near 3000 Greek, Roman; large, middle, and small Brass, and Denarii, in series; a superb collection of English Silver in fine preservation; several hundred Town-pieces and Tokens, Medallions in Silver and Brass of exquisite workmanship, by Dassier, Hammerani, &c.

The Fossils comprehend native and extraneous petrifactions, chrystals, marquisetts, spars, silver, copper, tin, lead, and iron ores; some specimens of porphyries, granites, &c. Shells for the cabinet or grotto; a repository or book-case, handsome mahogany medal cabinets, nests of drawers; a large iron chest, &c.

The whole 'tis presumed forms a Collection superior to any of the kind ever offered for sale in Bath.

To be viewed on Monday and Tuesday preceding the sale, which will begin each morning at eleven o'clock.

Catalogues to be had at the place of sale, and of W. Cross, upholder and undertaker, in Milsom-street; and of Messrs. Oliver and Ridout, the corner of High-street, Bristol. [2888

9 'All the pictures, prints, drawings, medals, coins, fossils and shells, the property of Mr. Hamlyn', Advertisement, Bath Chronicle, 8 May 1783

three volume Outlines of the natural history of Great Britain (1769-71) with good coverage of geological material, and a revised edition came out after his Bath period in 1788.

Practical geology

One field of geological communication was certainly germane to the development of the science in the area - the field of economic geology as we would call it today, i.e. the activities of the so-called 'practical men'. Such men had a long association with three geological facets of the area: the Bath stone industry, the associated Bath building industry and the nearby Somerset coalfield centred on Radstock to the south of Bath, whence came the coal used in the Bath market. (126)

Prospecting for Bath stone or for coal in the Somerset coalfield did not present great problems in this period. In the Somerset coalfield there was already a considerable tradition of surveying skills which could be called on, (127) as William Smith did soon after his arrival in the Somerset coalfield. (128) But practical men were *also* active in areas far outside the coalfields in the search for further deposits of coal needed to fuel the growing industrialisation of the country. Records of the many attempts made to find coal outside the coalfields are not likely to be usually available for the simple reason people would not chronicle futile attempts after their expensive failure had become known. John Farey wrote in 1807 (129) of hundreds of instances, known to himself or William Smith, of such futile attempts to find coal in the southern and eastern counties of England where none was to be found. I shall describe one such case for the light it sheds on the work of such coal hunters and 'practical men' in the days before scientific prospecting methods were developed. (130)

Attempts to find coal near Bath in the eighteenth century were very few because the supply from the nearby coalfield was able to meet the demand. Farther inland in South Somerset and Dorset, coal was much more expensive because of the added and considerable cost of carriage from the mines and it was here that attempts were frequent. The best documented attempt known to me was at Shaftesbury in Dorset, twenty miles south of the coalfield, in 1791, the year of William Smith's arrival in Somerset. (131) Shaftesbury was then a thriving county town with a population of over two thousand. It had long felt the need for cheaper coal; the first recorded attempts to find it were in about 1690 when specimens of the Upper Kimmeridgian shale, in which the attempt was made, yielding diagnostic ammonites and a large pliosaur vertebra, were found and passed into John Woodward's collection where they are still preserved today in Cambridge. (132) The attempt was made to the south of the town, which is built on Upper Greensand unconformably overlying the Kimmeridge clay which superficially resembles, and was many times mistaken for, the Coal Measures. Attempts were made in this Kimmeridge clay again in 1791; we first hear of this adventure in a Salisbury newspaper notice of December 1790 (133) which reported 'so great an appearance [of coal] as to afford well-founded hopes of success' and that other attempts had also been made about sixty years previously.

The 1791 adventurers were inspired especially by two events. Firstly by other attempts in 1790 farther afield; one at Chard in Devon on the Lower Lias shales there which inevitably came to nothing but which did not stop further attempts in the very same place in 1826, the other in Hampshire under the management of a Mr Jeffery of Salisbury, thought to have been Henry Jeffrey (c.1767-1819) a chemist there. The site of this attempt is not known but seems likely to have been the trial, personally observed by William Smith in the spring of this same year opposite the 'Shoe Ale-House' at Plaitford (134) only eleven miles south-east of Salisbury, in the much more recent Bagshot Beds of the Tertiary, and was equally unsuccessful.

The Shaftesbury attempt was also inspired by the advice of a famous, though now little known, itinerant science lecturer, John Warltire (c.1739-1810), (135) who lectured much at Bath (136) and in the West Country, specifically at Salisbury in 1775 and 1786. (137) Indeed, one of his 1786 lectures dealt with the formation of minerals and petrifactions (see Illustration 10). He gave as his opinion in 1786, according to two of the 1791 adventurers who must have attended his lectures, that there was 'one continuous bed of coal from Henstridge Ash [in Somerset] to Sherborne [in Dorset]', eight miles apart. (138) He was probably referring to the arcuate and faulted outcrop of Forest Marble of Middle Jurassic age, between these places which many people before and after were also to confuse with true Coal Measures. John Warltire was of Greek descent and had settled in England by 1762. He quickly became a sought after and popular itinerant lecturer in natural philosophy and his help was acknowledged by Joseph Priestley in isolating oxygen in 1774. Erasmus Darwin(1731-1802) invited his help about the same time and Warltire instructed both Darwin's and Josiah Wedgwood's children in science. He was elected an early honorary member of the Manchester Literary and Philosophical Society in 1782, and by the time of his visit to Salisbury he was at the height of his reputation. White Watson (1760-1835), mineral dealer of Bakewell, attended his lectures there in 1781 when he specifically lectured on mineralogy, and Watson also obtained chemical analyses of rocks from him. (139) William Gregor (1761-1817)

MR. WARLTIRE presents his compliments to the Ladies and Gentlemen of the City, Close, and Neighbourhood of SALISBURY, informs them that he intends opening his Course of Twelve Lectures, at the Parade Coffee-House, on Monday the 13th of November, at half after six in the evening, on the following subjects:

The Properties and Uses of the COMMON Air.

The Nature of the PRESSURE of Fluids, construction of various Machines for raising, and advantageously applying Water, &c.

The Doctrine of MECHANICS, and the application of that Branch of Science to the raising and moving heavy Bodies, &c.

The Science of OPTICS, or the application of different Glasses to form the various kinds of Microscopes, Telescopes, &c.

The construction of the Natural Eye, illustrated by Dissection.

The Discovery, Progress, present State, Power, Uses, &c. of ELECTRICITY.

GEOGRAPHY and ASTRONOMY.—Many new Ideas applied to explain Branches of that Science hitherto not understood.

The Method of procuring and applying to useful Purposes MANY KINDS of Air—being new Discoveries.

The Theory of the Earth, applied to account for the Formation, Change in Appearance, Alteration of the Properties, &c. of MINERALS and PETRIFACTIONS.

The Nature and Effects of HEAT, LIGHT, PHLOGISTON, &c.

The Utility of Mechanical and Chemical Philosophy to Mankind in general explained.

Mr. Warltire assures the Ladies, that NATURAL Philosophy, as now taught, is at once instructive and entertaining; every opinion being illustrated by experiment—which he is enabled to do, having in his possession a very extensive and elegant Apparatus, consisting of an ORRERY of a very complete kind,—GLOBES much improved,—AIR-PUMPS with the latest improvements,—MODELS of MECHANICAL and HYDRAULIC Machines,—a complete collection of OPTICAL INSTRUMENTS,—an extensive ELECTRICAL APPARATUS,—every thing necessary for the Chemical Part,—a curious Philosophical Collection of MINERALS, &c. &c.

SUBSCRIPTION ONE GUINEA.

For the accommodation of Ladies and Gentlemen from the country, and such as cannot make it convenient to attend in the Evening, a Morning Course as above will commence on Thursday the 16th inst. at 12 o'clock at Noon. [769

10 Announcement of lectures on natural philosophy by John Warltire, Salisbury and Winchester Journal, 13 November 1786

was inspired with a taste for mineralogy from his lectures in Bristol (140) and John Farey (1766-1826), the geologist, also refers to him with approval. (141) He can be regarded with Arden as representing the best in eighteenth-century itinerant science lecturers.

We can see then there were good grounds from a contemporary viewpoint for another attempt for coal near Shaftesbury. The 1791 attempt was made by a group limited to twenty local gentlemen and tradesmen who all initially agreed to subscribe £20 each to finance the attempt. In the available manuscript sources their names are merely listed, but it has proved possible to identify the majority of the adventurers. (142) The secretary was Charles Bowles (1766-1837), a Shaftesbury solicitor and brother of fellow adventurer, Rev. William Lisle Bowles (1762-1850), cleric and poet, the best known of the twenty and the only one to achieve an entry in the Dictionary of national biography. The Treasurer was the local Shaftesbury banker, Edmund Ogden (c.1748-1812), who is two years later found on the Dorset and Somerset Canal Committee. Apart from these, we find two described as gentlemen, including the Mayor, two clergymen apart from W. L. Bowles (all three were Oxford trained), one tailor-cum-publican, one other solicitor, one other publican, two grocers, two tanners, one edge-tool maker who supplied ironmongery and expertise for the attempt, one brewer, one surgeon, one post-horse supplier, and one whom is difficult to categorise called Lawson Hudleston (c.1745-1811) who, after Oxford University, had become an Indian civil servant and was highly regarded as an inventor and patentee. The chairman and the man largely responsible was another local surgeon called Richard Pew (1752-1834), MD St Andrews 1804. He is in many ways the most interesting of the group, having been trained at Edinburgh University (1775-7), where he became a Fellow of the Royal Medical Society and a friend of Joseph Black (1728-1799) who (by his lectures) interested Pew in chemistry and geology and with whom Pew later corresponded. (143) Pew published a chemical pamphlet in 1796 acknowledging his debt to Black. Pew had travelled on the continent as physician to the Bristol merchant James Ireland (c.1725-1814) before settling at Shaftesbury and this explains Pew's connections with Bristol and Bath circles and his election as an honorary member of the Bath Agricultural Society in 1792.

Another subscriber with Edinburgh connections was the surgeon and Mayor in 1790, John White (died 1793), on whose estate the main attempt was made. His son, John White junior (1767-c.1808), was in 1789 a medical student at Edinburgh University where he attended John Walker's (1731-1808) course on natural history. (144) We can see from Scott's edition of Walker's lecture notes dating from circa 1788 what White would have been introduced to on the subject of strata and geology. (145)

Another of the adventurers whose education is also likely to have encompassed some geology was the treasurer, Edmund Ogden. He was born in Liverpool about 1748 and in 1764, at the age of sixteen, entered the well known Dissenters' Academy at Warrington. (146) He would have been taught natural philosophy by John Holt (died 1772) and if he remained a student until 1767 he would have had the opportunity of attending John Reinhold Forster's (1729-1798) lectures on natural history and mineralogy given from 1767 to 1769. (147) Ogden had moved to Shaftesbury by 1772. Ogden also recorded receiving help with the attempt to find coal from a 'Mr Bright of Bristol' in April 1791 when he also asked Bright to send two men from Bristol to supervise the boring operation. These duly arrived. Mr Bright must be Richard Bright senior of Bristol (1754-1840), fellow banker and fellow Warrington Academy student (entered 1769). Bright and Ogden were related by both families having married into the Heywood family - a Liverpool banking family which included the treasurer of the Warrington Academy from its foundation. Bright was an early devotee of both geology and mineralogy, (148) and in 1780 he was elected an honorary member of the Chapter House Philosophical Society in London (149) whose members included Richard Kirwan (1733-1812), John Whitehurst (1713-1788), and many other eminent scientists, (150) such as James Watt, Josiah Wedgwood senior and Joseph Priestley, all with an active interest in geology and mining operations.

Backed by all this range of contemporary skills, advice and experience the team of twenty was convinced that Coal Measures were to be found on the outcrop of the Kimmeridge clay near Shaftesbury, and it made three different attempts to find them. The first went down to a depth of 405 feet using the same boring irons as the Hampshire attempt of 1790, before being abandoned. Surveyors and miners from Salisbury, Wells and Bristol were all involved. But the search proved quite futile and was abandoned in 1792. Another similar venture started in 1804, this time in the hope that the Oxford clay at Brewham near Bruton, a few miles north in Somerset, would yield coal. This venture cost a good deal more - £2750 - and reached 652 feet depth. (151) It is an identical story of misapplication with some of the same names, like Pew and Richard Messiter (1759-1830), a Wincanton solicitor, involved in this as in the Shaftesbury trials. The numerous attempts to find coal in the Jurassic and higher parts of the stratigraphic column in England in the last part of the eighteenth and early parts of the nineteenth century necessitate rejection of Rachel Laudan's claim that the ability to identify strata was 'commonplace' at this time. (152)

Summary: The nature of late eighteenth-century geological activity in Bath

In a notably hostile review of R.E. Schofield's book on the Lunar

Society of Birmingham, D. W. F. Hardie made the valid point that 'in democratic societies at least to assert guilt by association is generally considered bad law; finding significance by association and association alone, is equally bad historical procedure'. (153) Similar problems of association are raised on another front by Paul Kaufman who, writing of eighteenth-century reading and book-borrowing habits, noted that 'we cannot ever assume how much the library borrower anywhere did actually read - if he read - without some evidence'. (154) In short, even if a man borrowed or bought a particular book, we cannot from this be sure he read it! While in the case of C. H. Parry's ownership of John Walcott's 1779 book on Bath fossils we _can_ prove Parry was directly influenced by it, (155) we are much less able to _prove_ the significance of the association of different men who worked at the geology of Bath in the late eighteenth century. However, if we were to take D. W. F. Hardie too seriously, we would never attempt even to demonstrate the facts of such associations at all. This is all that has been attempted here. Peter Medawar has said that 'the factual burden of a science varies inversely with its degree of maturity' and this is also true of the history of science which is still at a very immature stage in its development. (156) This paper is purely a preliminary attempt to document some of the basic factual data in the development of eighteenth-century geology, as it applies to a particular but significant English city.

In general it can surely be shown that there _was_ a considerable amount of geological activity in this period in Bath, and that this activity has previously been almost entirely ignored by writers like Barbeau who have concentrated on the social and literary activity. (157) One can also demonstrate that, via such people as Edmund Rack and his fellow members of the first Bath Philosophical Society, there was considerable communication of that activity to the regions outside Bath both during the life of the Society and after. It is worth at least noting the wide social composition of the membership of the Philosophical Society, ranging in religious persuasion from Quaker, lapsed Catholic, Presbyterian, Methodist, both Wesleyan and Huntingtonian, to Church of England, and in background from tradesmen (stone carver, optician, soap boiler, coal merchant, etc) to clergy, itinerant lecturers, gentlemen, and physicians and surgeons. Medical men comprise ten out of the twenty-nine known members, and physicians with medical degrees eight out of the twenty-seven identified members. Of these eight, six were at least partly educated at Edinburgh, with medical training at Leyden, Oxford, Cambridge and Glasgow also apparent. The contribution of the medical men was a significant one. Also of significance is the fact that of the twenty-nine, fourteen are mentioned in the _Dictionary of national biography_, twelve were elected Fellows of the Royal Society, and of the remainder some, like John Walcott, at least deserved - but never got - notice in the D.N.B. One very important point about the nature of eighteenth-century Bath society in general has already been well made by Roy Porter. (158)

This is the lack of an indigenous culture there. The many visitors were attracted to Bath for particular reasons and the tradesmen, physicians and others who settled there, often for short periods, were all drawn to Bath from elsewhere for related reasons. This is shown by the fact that only two of the twenty-nine Bath Philosophical Society members (and one not positively identified) can claim to have been born in Bath. All the others whose places of birth are known were born elsewhere. This essentially peripatetic Bath culture was thus one singularly lacking in roots.

William Smith's contribution

One of the reasons this paper was undertaken was to examine Smith's debts to previous workers on Bath geology. To extend the analogy, if Bath was the cradle of modern English geology, was Smith the father or merely the midwife to previously conceived ideas?

William Smith (1769-1839) arrived in High Littleton, seven to eight miles from Bath, in 1791. By 1817 we have his acknowledgement of one obvious and direct connection with this previous activity in his use of John Walcott's book of 1779, but it is not known when Smith first encountered this book. (159) The man Smith himself mentions (160) as the first interested in his geological activities was Thomas Davis (c.1749-1807) of Horningsham, Wiltshire, agent and land steward of the Marquis of Bath's estates at Longleat where he served from about 1763 to 1807. (161) Davis was also in direct contact by 1800 (162) with the one man who might have directly stimulated Smith's interest in fossils, James Stephens (c.1748-1816), squire of Camerton near Bath, chairman and one of the major promoters of the Somerset Coal Canal, a major colliery owner and the man 'on whose estates [in 1796] Smith first put in practice his ideas of draining derived from a knowledge of the strata', (163) while he was still employed by the Canal Company. As we have seen, Stephens was a keen naturalist, a collector and student of fossils, and he had been elected an honorary member of the Society for Promoting Natural History in London in 1784. How, and if, Stephens did influence Smith's geology is not yet known.

Smith was elected a member of the Bath Agricultural Society on 13 December 1796, (164) probably supported by the next of his early contacts, the Rev. Benjamin Richardson (c.1758-1832), with whom Smith was seemingly in contact by 1797 (165) having been introduced to him by Davis. Even before his election to this Society, Smith's influence on Bath geological communication may have been felt. In April 1796 a letter appeared in the Monthly Magazine dated 17 March and written by John Hodder Moggridge (1771-1834) of Stokehouse, Bradford-on-Avon near

Bath, describing a 'Wonderful Phenomenon in Mineralogy'. (166) This was an illustrated description of the massive current bedding found in the Bathonian Combe Down Oolite at Bath of which the writer sought an explanation. (167) One could be forgiven for regarding this as evidence of an interest in the local geology quite independent of Smith. John Moggridge senior (1731-1803) was a clothier from Bradford-on-Avon and from 1794 Lord of the Manor of Dymock, Gloucestershire, (168) and he was also much involved with his son, from the end of 1797 at least, in coal mining operations, in the Newent coalfield, with Richard Perkins and his son of Oakhill, Somerset, who had been involved in this coalfield since October 1795. (169) Richard Perkins junior (c.1773-1850) was the son of another Richard Perkins (?1753-1821) who was a surgeon in Oakhill. Perkins senior had accompanied Samborne Palmer (1758-1814) of Timsbury, another colliery owner, and William Smith on their famous fact-finding mission for the Somerset Coal Company management committee to the North of England, Shropshire and Wales in August and September 1794. (170) Since, in November 1794, just after their return, Richard Perkins junior had married John Hodder Moggridge's sister Elisabeth at Dymock, (171) we can see that the independence of John Moggridge's geological interests from those of William Smith now need to be proved rather than assumed, for it seems quite likely Smith was the catalyst for the twenty-five year old Moggridge junior's interest in Bath geology. (172)

In June 1799 Smith was introduced to the Rev. Joseph Townsend (1739-1816), Rector of Pewsey, whose association with Smith is well known. One aspect of this association has however not been previously reported and may well have had a considerable influence on the early diffusion of Smith's ideas both in and around Bath. In late 1798 the second Bath Literary and Philosophical Society was instituted. (173) Records of this Society are even fewer than those of the first, but it has so far proved possible to identify ten members of it, including three who had belonged to the first Society. Sir William Watson, who had been such an important member of the old Society, was one of them and he seems to have played a leading role in the new Society. The member, however, with major geological interests was Joseph Townsend, who was certainly a member by late 1804. (174)

About the same time as Smith met Townsend, he also came into contact with William Cunnington of Heytesbury near Warminster (1754-1810), and in the next year Smith presented his geological results to the Bath Agricultural Society which voted him public thanks at the General Meeting on 10 June 1800, at which he was present, 'for his communication and observations relating to some new Improvements in Agriculture and Mining at the same time requests him to favor the Society with his further remarks on the subject'. (175) Two days later the newly elected secretary of the Society, Nehemiah Bartley, wrote on behalf of the Society to Smith 'thanking him for his observations on the various strata and to request further communications'. (176) A further connection with Bartley is that his third son, Thomas (1780-1819), previously a canal agent with the Kennet and Avon Canal, became for a short time clerk to Smith about 1800. (177)

It seems clear that Smith's total isolation from scientific contact was hardly a reality after his election to the Bath Agricultural Society. Just as Herschel was greatly helped by the activities of the Bath Philosophical Society, three of its members supporting his election as Fellow of the Royal Society, so it seems that Smith was similarly aided by the activities of the Bath Agricultural Society a few years later. Edmund Rack's two creations had certainly justified themselves.

Of the innovative nature of Smith's geological work there is very clear testimony. Using Walcott's 1779 book and Rack's manuscript notes on the 'Natural history of Somerset' of about 1785, (178) as barometers of the state of Bath geology just before Smith arrived in Somerset, we can see how clearly original and rapid his advances were. Both Walcott and Rack were mainly interested in fossils and minerals as collectors, of whom there were then a large number in the Bath area, (179) no doubt a direct result of the abundantly fossiliferous nature of the area, but neither paid any significant attention to the vital recording of localities of specimens or to the stratification of rocks.

The most eloquent testimony to the nature of Smith's achievement comes from those with whom he came into contact before the close of the century, such as William Cunnington, who wrote to Sir Richard Colt Hoare in 1809 in a letter describing the distribution of rocks and fossils along the Swindon to Chipping Norton road that 'the longer I live the more I am convinced of the importance of Smith's system [of identifying strata] and more heartily do I wish him to lay it before the public for examination'. (180) Similar comments came from Joseph Townsend, as quoted by Richardson:

> We [i.e. Townsend and Richardson about 1800] were soon much more astonished by proofs of [Smith's] own collecting that whatever stratum was found in any part of England the same remains would be found in it and no other. Mr Townsend _who had pursued the subject 40 or 50 years_ and had travelled over the greater part of civilised Europe declared it perfectly unknown to all his acquaintance. (181)

The coal hunters also bear passive but eloquent testimony to this originality. The attempt for coal in the Oxford clay at Brewham, near Bruton, was one which Smith personally warned against in 1805, and his originality in mineral surveying and prospecting using his knowledge of the strata cannot be challenged. Dr Richard Pew who, as already described, was involved in both the 1791 Shaftesbury attempt and the 1804 Bruton attempt was to write in 1819, nearly thirty years after his first involvement in coal hunting, that

> Mr [Robert] Bakewell's Elements [sic] of Geology [1813] or Mr Smith's Maps of the strata [1815] I think will throw a doubt upon the courage and expectation of finding coals here about [Shaftesbury], nevertheless from the strong <u>seeming</u> indications of coal between the bottom of East Stour Hill [Corallian Beds] and Toomer Hill [Forest Marble], Lord Digby, the Marquis of Anglesey and Sir Rich[d]. Hoare did at my suggestion about [1816] employ a Mr Bailey a miner from Staffordshire to make a survey. (182)

Despite Smith, Pew was still not finally aware of the reliability of prospecting techniques, for this part of the geological column, of which he had been totally ignorant before 1805. (183)

We may leave the final word on Smith to one man well able to judge, the Quaker William Matthews (1747-1816), who settled in Bath in 1777 as a brewer and coal merchant. (184) He was a founder member of both the first Bath Philosophical Society and the Agricultural Society of which he was also the second and highly successful secretary from 1787 to 1800, taking over from Edmund Rack. (185) He was thus in a well positioned vantage point for judging Smith's contribution to the development of geology in Bath. Early in 1800 he too called attention to Smith's novel discoveries, in a work published in Bath, writing thus:

> Then, hail, ye patriots, who can prize renown
> On peaceful plains, where sylvan honours crown!
> Let morning air invigorate your pow'rs,
> And plans of increase mark your passing hours.
> Waste long-neglected acres be your care,
> And barren wilds with fruitful fields compare.
> The depth and kind of surface-soil explore,
> How mark'd with fatness, or how tinge'd-with ore;*

> *Mr SMITH, an ingenious land-surveyor, of Midford, near Bath, has studied with success this subject, and the publick, it is hoped, will ere long be benefited by a publication of his important discoveries respecting the general laws of nature in the arrangement and external signs of under-strata, inclusive of fossils and minerals. (186)

This is a clear acknowledgement of the originality of Smith's skill at identifying the strata of Bath.

Notes

In collecting material for a study of this nature one quickly incurs numerous debts to the many librarians and archivists who so courteously look after and aid access to the collections in their care. My debts to these people are so many that I cannot name them here but must single out the three institutions in Bath itself without which this paper could not have been written and thank especially John Kite and all the staff of the Bath Reference Library, Bob Bryant of the Bath Record Office and Philip Bryant of Bath University Library, as well as my own University Library and archivist (Ian Fraser) for similarly valued help. Joan Cliff did wonders in typing my illegible manuscript.

For stimulating discussion and help on specific points I also thank Victor Adams, Robin Atthill, Jim Bennett, David Bick, Gavin Bridson, Warren Derry, Desmond Donovan, Joan Eyles, D. G. Hickley, Hal Moggridge, Roy Porter, Ian Rolfe, R. E. Schofield, John Thackray, Anthony Turner, and Michael Walcot.

1 J. Nichols, "Illustrations of the literary history of the eighteenth century", vol. iv, London, 1822, p.769.

2 At the conference in 1977 out of which this volume arose, Alex Keller discussed the carefully preserved multi-volume correspondence of the naturalist and fossilologist, E. M. da Costà(1717-1791). He rightly claimed this allowed, because of its complete nature, a study of those interested in fossils at this time as a group, and not just as individuals. This paper has the same aims for much of the same period but is framed instead round one place, Bath in Somerset (now Avon). Any study of this type is fraught with pit-falls, because of the incomplete nature of the material on which it is based, which was always dispersed and much of it since lost (or unknown to me), unlike the da Costa archive, which has survived intact as one accumulation. However, it would be wrong to assume even the da Costa archive was an entirely complete record of fossilology of the period. Josiah Wedgwood senior wrote of da Costa in 1774 to Thomas Bentley thus: 'Doctor [Thomas] Percival has sent me the famous naturalist Da Costa with injunctions to be very civil to him - I gain'd a little relief by sending him 2 miles to see a Flintmill but that is over, and now I am oblig'd to be rude to him whilst I write' (Wedgwood to Bentley, 7 August 1774, Wedgwood Archives Keele University 18551-25); and again ten days later: 'I do not know what character da Costa has . . . However he is gone. I left him on Tuesday, and he left the country on Wednesday. Dr P_____l is very high in his incomiums of da C. as a very sensible man of the most extensive knowledge and equally extensive correspondence with the literati all over Europe, amongst whom the Doctor says he is very much esteem'd. I thought him the most

disagreeable mortal who bore the name of Philosopher that I had ever known - Or I should not have left him so soon' (Ibid., [16 August 1774], 18552-25). One should not be surprised after this to find, as we do, only letters from Percival to da Costa and not from Wedgwood (who was of course deeply interested in geological matters) indexed in the da Costa archive.

3 A. J. Turner, "Science and music in eighteenth-century Bath", Bath, 1977, p.15.

4 Thus baptised by J. Hunter, "The connection of Bath with the literature and science of England", Bath, 1827, p.14.

5 J. H. Plumb, "The commercialisation of leisure in eighteenth-century England", Reading, 1973, p.20.

6 J. Britton, "The history and antiquities of Bath Abbey Church", London, 1825, pp.148-51.

7 "The universal British directory of trade, commerce and manufacture", Vol. ii, London, 1793, p.89. (Compiled c.1791.)

8 A. Barbeau, "Une ville d'eaux anglaise au XVIIIe siècle. La société élégante et littéraire à Bath sous la reine Anne et sous les Georges", Paris, 1904. (English edn.,1904.)

9 I. Woodfield, "The celebrated quarrel between Thomas Linley (senior) and William Herschel: an episode in the musical life of 18th century Bath", Bath, 1977; Turner, op.cit. (3).

10 A. Hare "Theatre Royal, Bath: the Orchard Street calendar 1750-1805", Bath, 1977.

11 H. R. Plomer et al., "A dictionary of printers and booksellers who were at work in England from 1726-1775", Oxford, 1932; lists twenty-seven booksellers active in this period in Bath. G. Pollard, "The earliest directory of the book trade by John Pendred 1785", London, 1955; adds several more active at this date. For surviving records of some of the circulating libraries run by these men see P. Kaufman, 'The community library: a chapter in English social history', Trans. Amer. Phil. Soc., 1967, n.s. lvii,part 7, pp.20-21, 62.

12 Barbeau, op.cit. (8), p.114.

13 Turner, op.cit. (3), pp.81-95.

14 R. Warner, "The history of Bath", Bath, 1801, p.344.

15 R. Porter, 'Metropolis, enlightenment and provincial culture: the social setting of Herschel's work', unpublished transcript, lecture given in Bath 1977.

16 Plumb, op.cit. (5), p.3.

17 R. S. Neale, 'The industries of the city of Bath in the first half of the nineteenth century', Proc. Somerset Archaeol. Natur. Hist. Soc., 1964, cviii, 132-44; see also R. A. Buchanan, "The industrial archaeology of Bath", Bath, 1969.

18 H. S. Torrens, "The evolution of a family firm: Stothert & Pitt of Bath", Bath, 1978.

19 W. Ison, "The Georgian buildings of Bath from 1700 to 1830", London, 1948.

20 J. J. Cartwright (ed.), 'The travels through England of Dr

Richard Pococke . . . during 1750, 1751, and later years', Publ. Camden Soc., 1888, n.s. xlii, 154-8; K. Hudson, "The fashionable stone", Bath, 1971.

21 S. Shaw, "A tour to the West of England in 1788", London, 1789, p.294.

22 M. E. Jahn, 'John Woodward, Hans Sloane and Johann Gaspar Scheuchzer: a re-examination', J. Soc. Bibliogr. Natur. Hist., 1974, vii, 23-4.

23 J. Woodward, "An attempt towards a natural history of the fossils of England", Part 1, London, 1729, pp.154-5.

24 Monthly Mag., 1807, xxiii, 24; J. Britton, op.cit. (6), pp.88-91.

25 M. E. Rowbottom, 'John Theophilus Desaguliers (1683-1744)', Proc. Huguenot Soc., 1968, xxi, 196-218.

26 R. E. M. Peach, 'Historic houses in Bath and their associations', 2nd series, London & Bath, 1884, p.26.

27 A. Elton, 'The pre-history of railways', Proc. Somerset Archaeol. Natur. Hist. Soc., 1963, cvii, 31-59.

28 M. J. T. Lewis, "Early wooden railways", London, 1970, pp.251-77.

29 B. Boyce, 'Mr Pope in Bath, improves the design of his grotto', in C. Camden (ed.), "Restoration and 18th century literature", Chicago, 1963, pp.143-53.

30 J. Wood, "Description of Bath", (1st edn., 1749) 2nd edn., 2 vols., London, 1765, i, chapters VI, VII. (Facsimile reprint, Bath, 1969.)

31 Ibid., i, 55.

32 Ibid., i, 61.

33 Ibid., i, 63.

34 C. Lucas, "An essay on waters", 3 parts, London, 1756, iii, 235-8.

35 British Library Add. MSS. 28538, ff.45-65, dated 1753-62.

36 Thomas Pennant, 'Reliquiae diluvianae or a catalogue of such bodies as were deposited in the earth by the deluge', British Museum (Natur. Hist.) Palaeontology Library MSS., vol. iii, 339.

37 S. Savage, "Catalogue of the manuscripts in the library of the Linnean Society of London. Part IV. Calendar of the Ellis manuscripts", London, 1948, p.13.

38 Original portrait in the National Museum, Stockholm; see W. D. I. Rolfe, in J. M. Chalmers-Hunt (ed.), "Natural history auctions 1700-1972: a register of sales in the British Isles", London, 1976, p.33.

39 Bath Chronicle, 20 December 1770.

40 Nichols, op.cit. (1), pp.762-9.

41 Ibid., p.766.

42 Archives of Society of Antiquaries of London.

43 North Munster Antiq. J., 1945, iv, 124.

44 British Library Add. MSS. 29743, ff.64 & 66.

45 Public Record Office. Dublin. D 18681, 1 March 1759.

46 Notes Queries, 1896, 8th series, ix, 383.

47 British Library press-mark 824.b. 17 (6).
48 Transcript in Bath Reference Library.
49 Each part was one shilling; an original wrapper of part 13 is preserved in University College (Natural Sciences Library), London.
50 Letter from Darwin to Walcott, 3 January 1780, transcribed on front end paper of a copy of Walcott's, "Flora Britannica", in the E. Green Collection, Bristol City Reference Library.
51 London Mag., 1779, xlviii, 374.
52 W. Smith, "Stratigraphical system of organized fossils", London, 1817, p.v.
53 See note (47), lot 894.
54 Walcott quoted from Stillingfleet, (2nd edn., London, 1762, p.175) who himself was quoting from C. Gedner, "Of the use of curiosity".
55 Walcott, "Descriptions and figures of petrifactions . . . ", Bath, 1779, p.iii.
56 Ibid., pp.13, 30-1.
57 Ibid., figs. 46-7.
58 "Mineral Conchology", 7 vols., London, 1812-46, iv, 106, plate 377, fig.2.
59 A. Percival, 'Biographical note on Edward Jacob', in E. Jacob, "History of Faversham", reprint ed. J. Whyman, Sheerness, 1974, pp.55-61.
60 John Walcott senior also possessed a copy of Jacob, "History of Faversham"; see note (47), lot 716.
61 "Plantae Favershamienses [or] a catalogue of the more perfect plants growing spontaneously about Faversham in the County of Kent, etc.", London, 1777.
62 H. Rolleston, 'Caleb Hillier Parry', Ann. Med. Hist., 1925, vii, 205-15.
63 Britton, op.cit. (6), p.131.
64 C. H. Parry, 'Parry', in W. Macmichael (ed.), "Lives of British physicians", London, 1830, pp.275-304 (300).
65 An appeal in 1890 for information about them yielded nothing, nor has a recent appeal been successful; 'Gloucestrensis', 'Dr Parry's proposed history of Gloucestershire fossils', Gloucestershire Notes Queries, 1890, iv, 507-8.
66 R. Cleeveley, 'The Sowerbys, the Mineral Conchology and their fossil collection', J. Soc. Bibliogr. Natur. Hist., 1974, vi, 421; J. Britton (ed.), "Beauties of England and Wales", vol. xv, London, 1814, p.314.
67 In possession of the writer.
68 Charles Henry Parry autobiographical memoirs, Bodleian Library, Oxford, MSS. Eng. Mis. d. 613, p.204.
69 E. Rack, "A dissultory journal of events, etc. at Bath Dec. 22 1779 to March 22 1780", Bath Reference Library MSS. 1111.
70 T. Curtis to C. Blagden, Royal Society MSS. Blagden letters C. 135 (28 March 1783) and C.142 (no date, but soon after the last).

71 E.g. by J. M. Edmonds, 'The geological lecture-courses given in Yorkshire by William Smith and John Phillips 1824-1825', Proc. Yorkshire Geol. Soc., 1975, xl, 373.

72 K. Hudson, "The Bath & West - a bicentenary history", Bradford-on-Avon, 1976.

73 A preliminarystudy with a list of members and the printed rules of the Society are to be found in Turner, op.cit. (3). Two more members can now be added to the twenty-seven listed there.

74 Turner, op.cit. (3), p.88.

75 Priestley's previously unknown connection with the Society is given by Rack, op.cit. (69). He had also already been in contact with Joseph Townsend (1739-1816), later William Smith's friend; J. Priestley, "Experiments and observations relating to various branches of natural philosophy", vol. i, London, 1779, p.208.

76 J. L. E. Dreyer (ed.), "The scientific papers of Sir William Herschel", 2 vols, London, 1912, i, pp.lxv-cvi.

77 R. E. Schofield, "The Lunar Society of Birmingham", Oxford, 1963, p.189.

78 Phil. Trans., 1778, lxviii, 789-90,

79 C. C. Hankin, "Life of Mary Anne Schimmelpenninck", London, 1860, pp.89-91.

80 A. A. Lisney, "A bibliography of British Lepidoptera 1608-1799", London, 1960, pp.222-3.

81 Catherine Wright to William Withering, letters 1784-1787, Royal Society of Medicine Library, London.

82 Dobson's membership of the Bath Society is given by his obituary notice in the Bath Chronicle, 29 July 1784, probably written by Edmund Rack. I owe this reference to the kindness of Warren Derry.

83 R. Gunnis, "Dictionary of British sculptors 1660-1851", London, 1968, pp.292-3.

84 He is perhaps the P.T. who signs a letter to the Gentleman's Mag., 1788, lviii, 793, about the 'thousands of petrifications of once living animals' to be found in the excavations then in progress on the slopes of Lansdown.

85 Gentleman's Mag., 1779, xlix, 432-3.

86 Dreyer, op.cit. (76), pp.lxvi-lxvii.

87 Royal Society of London MSS. Letters & Papers VIII, no.8, 12p.

88 Bath Agricultural Society archives, vol.10, accounts for 1777 to 1796, Bath Record Office.

89 'Anecdotes of Mr Edmund Rack', Europ. Mag., 1782, i, 361.

90 E. Green, "Bibliotheca Somersetensis", 3 vols., Taunton, 1902, ii, 317.

91 Ibid., i, 429.

92 Smyth of Long Ashton MSS. (uncatalogued), Bristol Record Office.

93 R. T. Gunther, "Early science in Oxford", vol.xi, Oxford, 1937, p.370.

94 Bodleian Library. Ashmolean MSS. 1822, f.87.

95 "History of the collections contained in the Natural History Departments of the British Museum", 2 vols., London, 1904, i, 269-70.

96 W. H. Curtis, "Life of William Curtis", Winchester, 1941, pp.16-17.

97 R. T. Gunther, "Early science in Oxford", vol.iii, Oxford, 1925, p. 224; a reference I owe to the kindness of Joan Eyles.

98 J. Hunter, "Observations and reflections on geology", London, 1859, pp.xliii-xliv.

99 A. T. Gage, "A history of the Linnean Society of London", London, 1938, p.5. (The original MSS. of Rack's paper survive in the Society's archives.)

100 Turner, op.cit. (3), pp.83-6.

101 J. Priestley, "Experiments and observations relating to various branches of natural philosophy", vol. ii, Birmingham, 1781, pp.379-82.

102 Arden to Wedgwood, letters dated 1763-64, Wedgwood Archives, Keele University, (1-616/9, 1-30225/6).

103 Bath Chronicle, 29 December 1768, 1 March 1770.

104 T. F. Plowman, 'Edmund Rack', J. Bath West England Agr. Soc., 1914, 5th series, viii, (p.21 of offprint).

105 British Library Add. MSS. 28534, ff. 118-21.

106 V. A. Eyles, 'The extent of geological knowledge in the eighteenth century',in C. J. Schneer (ed.), "Toward a history of geology", Cambridge, Mass., 1969, pp.159-83 (175-9).

107 Gentleman's Mag., 1781, li, 617 and fig.1.

108 The Society archives are preserved in the Linnean Society Library, London.

109 See note (88).

110 M. L. Mare & W. H. Quarrell, "Lichtenberg's visits to England", reprint, New York, 1969, p.95.

111 For an estimate of some of the numbers of visitors to Bath in this period, see S. McIntyre, 'Towns as health and pleasure resorts - Bath, Scarborough and Weymouth 1700-1815', University of Oxford D.Phil. thesis, 1973, p.463.

112 Turner, op.cit. (3), p.95.

113 D. Lysons, "Collectanea", 5 vols., no place, n.d., (British Library press-mark 1889 c5).

114 Gentleman's Mag., 1784, liv, 485-6.

115 Wood & Cunningham (publishers), "The improved Bath guide; or picture of Bath and its environs", Bath, [1809], pp.78-9. For Rack, see note (69). William Smith's offices in Bath at 2 Trim Bridge (Torrens, op.cit. (18), pp.20, 22) from 1802 to 1805 contained his fossil collections on display but not, apparently, to the public; L. R. Cox, Proc. Geol. Ass., 1941, lii, 16.

116 E.g. 'Medals, coins, fossils, etc. to be sold . . .', Bath Chronicle, 2 January 1783; 'All the pictures, prints, drawings, medals, coins, fossils and shells, the property of Mr Hamlyn (who is going abroad)', Bath Chronicle, 8 May 1783.

Reproduced as Illustration 9. Mr Hamlyn is the miniature painter John Hamlyn mentioned above.

117 James is claimed as the original ascriber of the title 'The harmonious blacksmith' to Handel's well known composition which he supposedly published with this title. No copy survives to prove this, but a James Lintern, blacksmith, appears in the 1837 Bath directory as further evidence of a strangely triangular family business. E. Blom (ed.), "Grove's dictionary of music and musicians", London, 1954, v, 250.

118 E. Robinson, 'The Lunar Society: its membership and organisation', Trans. Newcomen Soc., 1964, xxxv, 153-77 (163).

119 J. A. de Luc, "Geological travels in some parts of France, Switzerland and Germany", 2 vols., London, 1813, ii, 368.

120 Bath Chronicle, 10 February 1785.

121 J. A. de Luc, "Geological Travels", 3 vols., London, 1810-11, ii, 206-13.

122 See notes (21) and (84).

123 Lord Cardross to Catcott, 17 October 1766, Catcott MSS., Bristol City Reference Library.

124 "Dictionary of national biography".

125 Berkenhout to Herschel, 1 February 1788, Herschel MSS. W13 B 60, Royal Astronomical Society Library, London.

126 J. A. Bulley, 'To Mendip for coal - a study of the Somerset Coalfield before 1830', Proc. Somerset Archaeol. Natur. Hist. Soc., 1953, xcvii, 46-78.

127 Ibid., pp. 63-5.

128 J. G. C. M. Fuller, 'The industrial basis of stratigraphy: John Strachey (1671-1743) and William Smith (1769-1839', Amer. Ass. Petrol. Geol. Bull., 1969, liii, 2272-3.

129 J. Farey, 'Coal', in A. Rees (ed.), "The cyclopaedia or universal dictionary of arts, sciences and literature", vol.viii, London, 1807.

130 A preliminary account has appeared in H. S. Torrens, 'Coal exploration in Dorset', Dorset Mag., 1975, no.44, 31-9.

131 Documented: (a) 'MSS. Shaftesbury occurrences - topography, miscellaneous 1820-1830 collected by John Rutter' at Shaftesbury Museum. This contains a collection of the original minutes, bills and printed ephemera relating to the 1791 trial.

(b) MSS. draft c.1827 of 'An historical and descriptive account of the town of Shaftesbury", by John Rutter, which was never published; Dorset Record Office D/50/1.

132 J. Woodward, "An attempt towards a natural history of the fossils of England", London, 1728, tome ii, 52, 99.

133 Salisbury and Winchester Journal, 20 December 1790.

134 J. Phillips, "Memoirs of William Smith Ll.D.", London, 1844, p.5.

135 D. McKie, 'Mr Warltire, a good chymist', Endeavour, 1951, x, 46-9; N. G. Coley, 'John Warltire 1738/9-1810 itinerant lecturer and chemist', West Midlands Stud., 1969, iii, 31-44.

136 Bath Chronicle, 21 September 1776, 21 March 1788.
137 Salisbury and Winchester Journal, 13 November 1775, 13 November 1786. For these references and note (133) I am greatly indebted to Victor Adams.
138 Richard Pew to John Rutter, 12 February 1819, Shaftesbury Museum, see note (131); Rev. W. Blandford to Miss C. Bower, 6 February 1791, Dorset Record Office KW8.
139 W. Watson MSS. 1803, Derbyshire Record Office 589, Z.Z.6; W. Watson, 'On Entrochal Marble', Bakewell, 1826, single sheet.
140 L. Trengrove, 'William Gregor (1761-1817) discoverer of titanium', Ann. Sci., 1972, xxix, 362.
141 J. Farey, 'Cursory geological observations lately made . . .', Phil. Mag., 1813, xlii, 58.
142 Drawing on a variety of sources but especially 1783, 1784 and 1791 Directory lists and the fine collection of Shaftesbury deeds preserved in the Dorset Record Office.
143 Pew to Black, 12 October 1784, Black MSS. ii, 228-9, Edinburgh University Library.
144 MS. list of students, at the class of natural history in the University of Edinburgh, 8 May 1789, Edinburgh University Library. I owe this information to the kindness of Joan Eyles.
145 H. W. Scott (ed.), "Lectures on geology by John Walker", Chicago, 1966.
146 J. F. Fulton, 'The Warrington Academy (1757-1786) and its influence upon medicine and science', Bull. Inst. Hist. Med., 1933, i, 50-80.
147 M. E. Hoare, 'Johann Reinhold Forster (1729-1798): problems and sources of biography', J. Soc. Bibliogr. Natur. Hist., 1971, vi, 4.
148 Buckland, obituary notice of Bright, in address to Geological Society, 1841, Proc. Geol. Soc. Lond., 1841, iii, 520-2.
149 Wedgwood archives, Keele University.
150 R. W. Corlass, 'A Philosophical Society of a century ago', Reliquary, 1878, xviii, 209-11.
151 G. Sale, "Four hundred years a school: a short history of King's School, Bruton", Bruton,n.d., unpaginated.
152 R. Laudan, 'William Smith. Stratigraphy without palaeontology', Centaurus, 1976, xx, 210-26 (225).
153 D. W. F. Hardie, [review of R. E. Schofield, "The Lunar Society of Birmingham"], Business History, 1966, viii, 73-4. See also M. Berman, [review of A. E. Musson & E. Robinson, "Science and technology in the Industrial Revolution"], J. Social Hist., 1972, v, where the same point is made.
154 P. Kaufman, "Libraries and their users", London, 1969, pp.34,88.
155 See note (68).
156 P. B. Medawar, "The art of the soluble", London, 1967, p.114.
157 See note (8).
158 Porter, op.cit. (15), p.11.
159 See note (52).

160 Wm. Smith MSS. 1831, Department of Geology, Oxford University.
161 Monumental inscription in Horningsham church by Henry Westmacott.
162 L. Blomefield, 'Copy of a letter from Mr Stephens of Camerton near Bath to Mr Davis of Longleat on the subject of diseases of wheat dated August 22 1800', Proc. Bath Antiq. Field Club, 1877, iii, 12-16.
163 W. Smith, "Observations on the utility, form and management of water meadows", Norwich, 1806, p.54; see Phillips, op.cit. (134), p.16.
164 Archives of Bath Agricultural Society, v, 182, Bath Record Office.
165 J. A. Douglas & L. R. Cox, 'An early list of strata by William Smith', Geol. Mag., 1949, lxxxvi, 180-88 (182).
166 Monthly Mag., 1796, i, 205.
167 Turner, op.cit. (3), pp.13-14.
168 J. E. Gethyn-Jones, "Dymock down the ages", privately published, 1951, p.30.
169 D. E. Bick, 'The Newent Coalfield', Gloucestershire Hist. Stud., 1971, v, 75-80; D. E. Bick, 'The Oxenhall branch of the Herefordshire and Gloucestershire Canal', J. Railway Canal Hist. Soc., 1972, xviii, 71-5.
170 Phillips, op.cit. (134), pp.6-14.
171 Gentleman's Mag., 1794, lxiv, 1054.
172 For valued help with genealogical information on the Moggridge and Perkins families, I gladly thank Hal Moggridge, Robin Atthill and the Gwent County Archivist.
173 F. Shum, "A catalogue of Bath books", Bath, 1913, p.20; J. Britton, 'Bath', in A. Rees (ed.), "The cyclopaedia or universal dictionary of arts, sciences and literature", vol.iii, London, 1803.
174 James Currie to William Roscoe, 7 February 1805, Roscoe MSS. 1108, Liverpool Reference Library.
175 Archives of Bath Agricultural Society, v, 10 June 1800, Bath Record Office.
176 Ibid., iii, 12 June 1800.
177 B. Richardson to W. Smith, 31 December 1804, Wm. Smith MSS., Department of Geology, Oxford University.
178 This consists of a separate notebook in Rack's hand listing clays, marls, ochres, tripelas, fossils (i.e. minerals), spars, crystals, stones, geodes, and fossil shells found in Somerset with an 'Account of the coal mines near Stowey and Faringdon and Chew Magna' which is almost entirely derived from John Strachey's pioneer work. These notes intended for and partly used in John Collinson's History of Somersetshire give no clue that Rack was aware that Somersetshire strata could be identified or tabulated in any way; MSS. at Bristol Record Office.
179 Phillips, op.cit. (134), pp.17, 28.

180 R. H. Cunnington, "From antiquary to archaeologist", Aylesbury, 1975, p.141.

181 B. Richardson, letter of 10 February 1831, in A. Sedgwick, address on announcing the first award of the Wollaston prize, Proc. Geol. Soc. Lond., 1831, i, 270-9; emphasis added.

182 See note (138).

183 There is a serious need for research into the evolution of coal prospecting techniques both inside and outside the known coalfields. If the experience from Somerset is typical, practical men can be demonstrated as often particularly bad at identifying strata. For the Bruton attempt of 1804-10, already twice noted, the sinker in charge of a notably abortive trial was Gregory Stock, mining agent of Ashwick, Somerset, who later appears as a source of stratigraphic information for the southern part of the Somerset coalfield supplied to W. Buckland and W. D. Conybeare, 'Observations on the South-Western coal district of England', Trans. Geol. Soc., 1831, 2nd series, i, 210-316 (270-1). The lunatic waste of money spent searching a second time for coal on Chard Common, Somerset, in 1826, went on employing Samuel Barton of Nottingham to bore at a point he chose himself on the Lower Lias shales where as he said 'the local Coal Measures appeared to be the most laid bare'! He bored a 7½ inch diameter hole 379 feet deep which failed to leave the Lias. (Somerset Record Office, DD/CN box 28, bundle 21.) Barton, a coal agent, was one of the men who helped John Farey with his survey of Derbyshire. See his "General view of the agriculture and minerals of Derbyshire", London, 1811, p.xvii; he might thus have known better.

184 Turner, op.cit. (3), p.90.

185 W. Matthews, "A dissertation on rural improvements", Bath, 1800, p.65.

Geological controversy and its historiography: the prehistory of the Geological Society of London

PAUL JULIAN WEINDLING

The genesis of geology as a modern science in England has conventionally been attributed to the foundation of the Geological Society in 1807. It was the first specialised society of that name. Members were pioneers of stratigraphical fieldwork. It was to be a centre of research and exchange of information. The Society has therefore been considered a major cause of the switch from speculative theories of the earth to a science grounded in empirical research. Fitton, an early stalwart, ascribed this to 'plain men who felt the importance of the science'. (1)

As in other sciences, its practitioners acted as the first historians. Their story, that the beginnings of modern geology were the internal history of the Society, has been deferentially accepted by professional historians of science. The latter have been concerned primarily not so much with chronicling particular discoveries, but with the general problem of when geology became a fully-fledged science. If some historians say this was not until Lyell in the 1830s, it was still within the confines of the Society. But criteria of what constituted a science have varied according to different periods and prejudices. The enterprise of trying to locate a single turning point, when science is viewed primarily as a linear and cumulative process, is not fully historical. It has resulted in excessively severe criticism of other efforts which were contemporary with the foundation of the Geological Society. Societies which were oriented in a practical direction, as in Newcastle and Cornwall, have been condemned as too partial in their scope, whereas the Royal Society was apparently too broad. Mineral surveyors have correspondingly been written off as too empirical, and mineral collectors as too speculative. They have been excluded from being recognised as legitimate members of the charmed social circle of truly 'scientific' geology. (2)

Historiography and mineral history

The Society's history has been seen in terms of unquestioned success - as if the Society had been formed by a catastrophist force of truth. Geikie's Centennial Address and Woodward's Centenary History were correct that the mineralogists and chemists who formed the Society had widening fields of observation due to increased travel, and that they were drawn together by London mineralogical cabinets. Both explained this as due to a transition from speculation to science. But there is a discrepancy in the explanation of a progressive accummulation of truth. The knowledge of many members left much to be desired, however flattering Fitton's excuse that 'the want of education is sometimes of advantage to a man of genius'. (3) Rather than historically analysing the immediate background of London mineralogy with its technical and socio-economic concerns, recent historians have preserved the purity of geology as a discrete scientific entity; by relating it to broader cultural factors like earlier traditions of inquiry, religion and 'taste', they have tried to demonstrate the superior objectivity of the Society. Martin Rudwick has described how the rival hypotheses of Hutton and Werner were tested by Greenough. (4) Greenough then became President of the Geological Society, founded in 1807, and forced the Society into independence from the hegemony of the Royal Society. The testimony of this ambitious social-climber has been used to generalise about the instituting of a new 'English school'. (5) But once the sequence of events that led to the foundation of the Society are examined, Greenough's account can be criticised as too biased to warrant such generalisations. Whereas Roy Porter has recognised that geology was the culmination of a long-term process of increasingly scientific conceptualisation of the earth, he neglects short-term factors such as exactly how the group of founders arose. Although he goes one step towards correcting the naive view regarding the novelty of the early Society, his critique does not go far enough because he still accepts that the Society at last practiced modern geological science. Recognising the problem created by those members more concerned with politeness than with knowledge, he has interpreted the Geological Society's success as expressing a new force for social status. It was an exclusive club of leisured gentlemen; science had created a self-sustaining social structure, in which to participate was a generally esteemed mark of distinction. (6) Its amateurism seems to contrast with the state mining schools of Freiberg, Schemnitz, and especially the Paris *École des mines* established in 1783. However, the connection between the foundation of the Geological Society and a preceding scheme for a national school of mines is examined here to show that a stimulus for certain practitioners of geology and mineralogy was that these sciences offered a means of social improvement.

The Society's aim was 'the advancement of Geological Science more particularly as concerned with the Mineral History of Great Britain'. (7) Rather than the establishment of any peculiarly English school in a methodological sense, what was intended was a national survey of the country's mineral wealth using Continental methods of research. This would enable the application of geology to economic purposes, and contribute to a theory of the earth based on particular local studies. (8) There has been disagreement as to whether geology should be seen at the peak of industrialisation, as an elite leisure activity, or at its base, as applied science. For Roy Porter, economic geology was marginal to the science's inherent social structure. (9) Maurice Berman has presented an alternative case: the geology and mineral collecting of Davy at the Royal Institution were dictated by the Managers' entrepreneurial ideology. (10) It is my contention that as far as mineral history motivated the Geological Society, it represents an economically-oriented factor in its foundation. Mineral history meant examining the internal chemical characters of minerals as well as their stratigraphical relations. As a result, techniques of crystallography and mapping were greatly improved. Though they were far from new in 1807, stratigraphical observations were used for drainage and canal building (although this has not been studied in detail using estate maps and industrial records). The considerable attention paid to crystallography at this time has largely been overlooked. Mineralogical and stratigraphical observations had featured in technological tours, which were an important link between economic and scientific concerns. Besides offering picturesque views, tours were a chance to survey natural productions both in their native state and in mining and metallurgy. Mineral history, justified in Baconian terms, was far from naive empiricism. It represented the attempt to resolve the debate between the classification of minerals by internal characters, exemplified by the Linnaean approach, and external characters, which Werner had derived from mining experience. The travels of those who were to found the Geological Society indicate an increasing interest in this mineral historical approach. Greenough journeyed in the company of Davy, who was originally commissioned to collect for the Royal Institution and Board of Agriculture. Members of the technically-concerned British Mineralogical Society, founded in 1799, like A. Aikin, Lowry, Babington and W. Phillips, pioneered field work, yet they remained interested in chemical analysis and crystallography long after the Geological Society was founded. (11)

Collectors of men

Traditional accounts of the Geological Society assumed it was intended to be of general economic benefit. Halévy saw its foundation as a notable example of a general expansion of science, generated by industrialisation. He observed that 'scientific theory was the offspring of industrial practice'. (12) This relationship has since been questioned.

Differences of interpretation can be solved by detailed case histories of how the Society actually was formed. (13) The scientific activities of the Hon. Charles Francis Greville (1749-1809) are a suitable example. Although Patron of the Geological Society, he sided with the Royal Society when they were in dispute. He therefore lost his place in history, as Horner, an early member of the Geological Society, remarked. (14) Yet in putting together what was regarded as the finest mineral collection of his age, Greville had established himself as a key organiser of science. James Watt described him as 'a man in power and an amateur of science'. (15) Greville had undertaken tours for 'occasional researches into geology', and corresponded with others on tour, such as Charles Hatchett in 1796. (16) This was part of an extensive national and international correspondence on mineral matters which he maintained in close association with Joseph Banks, President of the Royal Society. Although Greville published only minor contributions to debates on meteors and corundum, he was active in transmitting information and specimens, so assembling a collection which enabled others to undertake research.

For example, William Hamilton has previously been seen as an isolated pioneer of exact geological observation. Greville, as Hamilton's nephew, was responsible for communicating his uncle's antiquarian and scientific papers, keeping him abreast of the latest cultural affairs (and passing on Emma, his mistress!). They shared economic interests with estates in the area of Milford Haven, where Greville was determined to establish ship-yards, a port and manufacturing, helped by a mathematical school to 'reduce theory to practice and speculation to use'. Greville remarked that 'the whole of the success in collecting persons to build depends on my personal connexions'. (17) One connection was Greenough, who had also become an M.P. in 1807. He was referred to by Greville as 'my mineralogical friend', in the context of his piloting a bill concerning Milford through Parliament in 1808. (18) This was just before the quarrel between the Geological and Royal Societies erupted. Greville had been a Lord of Trade, and he keenly followed the activities of the East India Company (especially its surveys of natural productions in which minerals were to the fore). The career of such a collector raises the question of how science was connected to his economic concerns, and whether the attitudes involved in these were central to the rise of geology. (19)

Porter has observed that declarations regarding the utility of science may mean purely cultural as opposed to economic improvement. He also states that the <u>laissez-faire</u> structure of industrial development meant that manufacturers would not allow diffusion of applied science. (20) Greville's enthusiasm for national mineral surveys to be used for economic purposes can be traced to Baron Born's international society for the exchange of mining information, founded at Skleno near Schemnitz in 1786. Greville was an honorary member of this association: 'Struck with the shackles imposed on mineralogists by monopolizers of new and useful processes, they thought no method so effectual to break them, as forming a society, whose common labours should be directed to fix mining on its surest principles.' (21) Greville had purchased the nucleus of his collection from Born, an enlightened mining engineer, who was the first to use Lavoisierian principles in his influential mineral classifications. Greville and Born remained in close correspondence, exchanging minerals and technical information which were relayed to others, like Black in Edinburgh.

At the same time the problems of the necessary fact-finding tours were shown by the mineralogical and technological journey attempted by Baron von Stein in 1787. Greville and Banks tried to open doors to this technically-trained bureaucrat, then in charge of the Westphalian mines. Boulton, Watt, and their employee, Raspe (the translator of Born and author of _Baron Munchausen_), closed them, by accusing Stein of being an industrial spy. Manufacturing interests could indeed be in opposition to applied science. Boulton wrote that he could help Stein as a naturalist, philosopher and gentleman, but as an engineer he would lay obstacles in his way. (22) Such individualism could still allow science to be applied if not diffused. Yet it is also arguable whether there was a _laissez-faire_ economic structure at so early a point in the Industrial Revolution. Boulton and Watt were protectionist in that they were zealous defenders of a strong patent system. Certain landowners realised, under the pressure of the French Revolution, that barriers to the diffusion of science would be overcome and that general promotion of economically useful science would strengthen their position as the ruling class. Greville was of more use to later tourists. In 1803 he helped the French _Ingénieur des mines_, Bonnard, and another 'industrial spy', Svedenstierna, who said, Greville 'not only offered new introductions to the largest ironworks of the kingdom, but also gave me an extract of mineralogical notes, which he had collected on a tour of South Wales and part of Cornwall'. (23) This suggests that there is a fault in the argument that economic geology was merely marginal to the establishing of the science. For if its pursuit was a sustained cultural response to remedy the lack of applied science, then the very efforts to promote utility mean the science should be viewed as a product of industrialisation.

This connection between the desire for economic improvement and the institutionalisation of geology can certainly be established in Greville's case. As early as 1790 Greville wrote of his aim to make a complete mineral history of Britain. (24) This plan emerged in a more mature form in 1804 with the proposal for a 'National Collection and Office of Assay'. This was to have had 'Professors of Geology, Mineralogy and Metallurgy'. It was to have been a centralised research institution on the Continental pattern, 'in which the whole grounds of Mineralogy and Geology illustrated by specimens and authenticated by chemical analysis can be investigated'. This was to match a view of capital-intensive industrialisation:

> The mining concerns of these kingdoms are conducted with such advantages of Capital that no other country can show Philosophical and Mechanical Powers applied with so much effect as in our mines; and no other Country is so deficient in the means of rendering the knowledge of minerals accessible to persons desirous of instruction owing to the want of a fund, applicable by Mineralogists and Chemists to the establishment of a Public Collection of Minerals. (25)

The plan was backed by two other aristocratic collectors, John St Aubyn and Abraham Hume. St Aubyn,having extensive Cornish mining interests, had employed Raspe with Boulton and Watt for assaying. His collection was first arranged by Babington on Born's Lavoisierian principles, which Greville had recommended as of use in mining. (26) Hume was also noted as a connoisseur, having a rich collection of precious stones; he had extensive estates and represented the East India interest in Parliament. The scheme was proposed as an auxiliary establishment to the Royal Institution, where Greville's associate Hatchett was Secretary. Davy was building up a geographically arranged collection for economic purposes there. (27) This was also where members of the British Mineralogical Society, another precursor of the Geological Society, were active with lectures and laboratory experiments.

The idea of a National Collection was based on the idea of a national survey. The proposed models for this were joint papers in the *Philosophical Transactions* by Hatchett, Bournon and Chenevix. These mineralogists had analysed muriate of lead from Cornish mines, and also the stratigraphically interesting oxide, corundum, of use as an abrasive. Greville had acquired the latter mineral in the course of a sustained mineralogical correspondence with India, where he had sent a list of mineral and antiquarian queries. A typical specimen of the replies was by Dr Percy of Calcutta in 1782: 'I shall be able to collect much useful information relating to objects of natural history as well as of antiquities in the country, I am attempting to form a Society for that purpose.' (28) Warren Hastings wrote,

> A gentleman who attended the Lectures of Professor Black at Edinburgh is now employed by the Company in search of Mines. I have transmitted your commands to him, and enjoined his particular

> attention to them, and will do myself the pleasure to send you whatever he may have collected by the ships of the next season, with such other natural productions or works of art as I can collect from different parts of the country. (29)

It is mistaken to see such a collector as isolated in his cabinet; he was a middleman encouraging the fieldwork of finders of minerals - here lay the providence of commerce! - and also encouraging chemists to provide useful information. Such examples suggest that science was indeed being economically applied: the problem is to ascertain the extent of this. William Smith, for instance, said his discovery of the stratigraphical significance of fossils reached the East Indies by 1802. (30) Details of the National Collection were sent to Madras, Bombay, Ceylon, St Helena, and Canton, as 'on a scale worthy of the BRITISH EMPIRE . . . and from the wisdom and liberality of the EAST INDIA COMPANY, great and effectual assistance may be hoped'. (31) Berman has observed that this collecting reflected the mercantile interest among the Royal Institution Managers, just as mining was calculated to appeal to the predominant landed interest. (32) Besides such an explanation based on the internal structure of the Royal Institution, broader factors, such as the need to cultivate science in economically underdeveloped and politically unruly areas like Wales and the colonies, evidently stimulated the scheme in Greville's case.

The National Collection foundered on the rocks of the self-interest of the Royal Institution's proprietors. Yet they used its publicity to increase membership and they maintained an Office of Assay and a collection. For 'a general meeting having resolved the proprietors only should be admitted as patrons, several subscriptions have been in the course returned and others prevented'. Public benefit and an extension of scientific research were here in conflict with the private interests of a scientific organisation. (33) But as originally conceived, the National Collection had adapted the Continental form of the professional mining school to that of the English convention of amateur patronage of science. It was also important as a transition, in that it brought together the group which was to found the Geological Society. An essential part of the plan had been to secure employment for Count Bournon, an _émigré_, who arranged the collections of Greville, Hume and St Aubyn. Bournon was then the centre of a group of mineralogists, which included the aforementioned patrons. They met at Babington's in 1804 for instruction in mineralogy by Bournon, and in 1807 they financed the publication of a treatise by Bournon. According to Greenough, this group continued its meetings which culminated in the foundation of the Geological Society. Greenough's account (given as a Presidential Address in 1834), failed to mention the National Collection or that meetings were held as early as 1804, although he was active at the Royal Institution. He asserted that the Society 'was the effect of accident rather than design'. Martin Rudwick has added that 'the earliest phase of the Society is in fact rather obscure. What is clear is that it arose from some informal gathering of amateur mineralogists; but the exact nature of the gathering is uncertain.' (34) But it now seems that there was a grand design - the National Collection - which resulted in contacts between previously distinct groups. As its purpose was a professional establishment for applied geology, even though financed by voluntary subscription, this modifies the idea of the amateur informalism of subsequent meetings. When Bournon observed that their patriotism was an outstanding feature, he did not mean that their science represented a new insular type of geology, but that it was to be applied for social improvement. (35)

The National Collection had a lasting effect: mineralogical and metallurgical research was stimulated at the Royal Institution, and it brought many of the founders of the Geological Society closer together. The Royal Institution may be compared to the Geological Society: in both, hopes for the application of science to social improvement encountered opposition. Thomas Webster, who conceived of the Royal Institution as a mechanics' school, had been sacked. (Ironically, he later became Secretary of the Geological Society.) The founders of the Institution, of whom Banks was a leading spirit until his resignation in 1805, envisaged science as a form of philanthropy to improve the welfare of the people, so as to prevent sedition. By 1804 Banks wrote to Rumford that the Institution had 'fallen into the hands of the enemy and is now perverted to a hundred uses for which you and I never intended it'.

In reaction to the French Revolution's boost to scientific research and organisation, there was a difference of opinion about the extent to which it was necessary to impose stringent controls on the spread and uses of science. Banks and Greville were more demanding in this respect than the Royal Institution. They hoped that applied science would result in social progress strengthening Church and State. (36) Regarding the problem of utility of science, the emergence of new concepts of paternalism - in response to the democratic challenge-must betaken into account. Philanthropic and utilitarian applications demonstrated the ability of science to offer new techniques of social control. Hence the ambitious scale of schemes such as Greville's, harnessing science to industry to strengthen the existing social order. Banks was anxious to communicate with French scientists, but he prided himself that by 1804 the Royal Society had 'not one attending member who is at all addicted to Politicks' and that its subordinate societies were in order. (37) The British Mineralogical Society was committed to science for humanitarian ends, with its members active in Rumford soup charities, leading to surveys of poverty and to support for Joseph Lancaster's monitorial education system. Their attitude was largely apolitical, many being Quakers for whom socially applied science could be a conscious substitute for radical politics.

Subscribers to the National Collection fitted this socially progressive, but politically repressive mould. The Bishop of Durham had extensive industrial interests. Robert Clifford, a populariser of Barruel's theory of a _philosophe_ conspiracy was a Vice-President of the Society of Arts, which offered premiums for mineralogical maps. Henry Englefield was a Roman Catholic, an aristocratic landowner, and a pioneer of stratigraphy; his research into monastic architecture was in reaction to the de-christianisation in France. Thomas Coutts, the banker, and David Pike Watts, the wine merchant, were both noted philanthropists. (38) Bournon, at the centre of the scheme, was, like many _émigrés_, distinguished for his social mode of thought, as his enthusiasm for applied science showed. All these various groups shared a horror of French atheist and atomist materialism, from which geology had to be redeemed. David Knight has described Davy's vitalist opposition to Lavoisier's material elements and weights. Roy Porter has seen 'the deluge of religious and political counter-revolutionary feeling' forcing geologists into a posture of extreme empiricism. (39) This interpretation can be extended to provide positive motives for practising the science for the sake of utility. Men were driven to institutionalise geology not because the earth was in decay, but because they feared civil society was. This type of social concern has been obscured because the Geological Society changed in character and motivation. It became swamped by those seeking public esteem; Greenough referred to 'poets, statesmen, historians and warriors', 'whose main objects in life, if not alien were connected but remotely with those of our institution, conferred upon it, notwithstanding, by their enlightened encouragement important advantage'. (40) This raises the problem of the fate of the institutional structure for organised research necessary to apply science.

In order to clarify the role of Banks and Davy, with their subsequent hostility to the Geological Society, it is necessary to establish their position with regard to applied science at the Royal Institution which to some extent underwent a similar transformation. Berman has reinterpreted the Royal Institution by showing it to be an offshoot of the Board of Agriculture's aim of improving techniques, and of the Society for the Bettering of the Condition of the Poor's rural philanthropy. It is in this context that interest in agricultural chemistry began. But after 1801, Berman suggests, philanthropy withered away. The purpose of Davy's geology at the Royal Institution was therefore purely entrepreneurial: to serve the East India Company and the landowning interest in mining. (41) This distinction seems questionable. It is based on Berman's analysis that the proprietors of the Royal Institution (who were composed of two apparently distinct groups, the landed and the mercantile) fullydetermined the type of scientific activity. But the initiative for Davy's appointment at the Board of Agriculture was taken by Banks. He suggested both the subject of agricultural chemistry, as the Board had only conceived of mechanics as applied to agriculture, and Davy as lecturer, with whom he personally arranged the conditions of employment. This included the performance of soil analysis. The Board did not accept Bank's proposal for agricultural botany. This meant that its members had only a passively selective role, rather than the actively innovative role played by the scientific expert. Banks combined his knowledge with his social position to indicate in which directions the extension of activities would be most fruitful. The narrower structure imposed by the Board was evident in that arrangements for Davy's lectures to be given to the public were made only in 1806. (42) How research produced institutional growth was shown by the Board setting up a laboratory in 'a room below stairs' for his work. The importance of the innovating enthusiasm of an expert was also apparent in the proposal for mineralogical surveys, raised in 1795 in the presence of Kirwan at the Board. Even earlier, Baron Born had suggested the need to Sir John Sinclair, the Board's founder, for a Board of Mines. (43)

Greville's initiative in putting forward the idea of the National Collection also conforms to this pattern. Davy was far from subservient regarding the lengths to which he took mineralogy in his theoretical papers given to the Royal Society, in his genuine belief that progress achieved by science would overcome class divisions, and in his personal enthusiasm for geology. He was able to combine the latter with his interest in electro-chemistry, from which Berman says the Managers diverted him. (44) He was even prepared personally to subscribe to the National Collection. That such teaching institutions were still being proposed shows that minerally-minded men were unaffected by the suggested change in attitude regarding education and philanthropy. Davy's associates in the British Mineralogical Society enthusiastically supported Lancasterian education, and Greville conceived his philanthropic foundation of a mathematical school as the model to be adopted by every county. Even the Board of Agriculture ventured to propose experimental farms 'as a sort of Academy or College' in 1806. (45) Such science had a generalising force: increasing depth of knowledge of nature went with greater breadth of social organisation. This attitude predominated among those mineral historians who would found the Geological Society.

The Royal Society applies itself to geology - and to the Geological Society

Banks and Greville as patrons and Davy as a professional all acquired influential social positions through their science. But it is only by taking into consideration their desire to apply science as a palliative for social ills that their emphasis on practical research can be explained. If Greville used his social position to gain backers for his Milford Haven scheme, its nature as a strategically-important model community went far beyond personal profit. Greville persevered with his plans even though Hamilton incurred substantial losses. Banks was a mine and landowner in Derbyshire, but only by understanding his broader social views can the commitment of the Royal Society to useful science be accounted for. It is important to examine this, so as to refute the view that Greville and Banks were merely being autocratic to protect their personal status, and that Davy was purely jealous of Greenough's position as President, when they clashed with the Geological Society. The concept of 'geology' was considerably developed in the Royal Society. In 1800 Banks hoped that discovery of silver in Herland mine 'may be useful by exciting inquiries into the science of Geology'. (46) An example of this research programme was Hatchett's analysis of the Mere of Diss; he used observation of present processes to explain the past origins of pyrites in coal mines. In three further papers, he dealt with the origin and formation of various types of coal, providing an insight into fossilisation. These experiments yielded a tanning agent - a topic of interest to Davy as it related to his work for the Royal Institution. The chemical approach to an understanding of fossils was continued by later members of the Geological Society. Such research proves the difference between a scientific organisation which was run by men who were able actively to contribute to science, like Banks and Greville, who was Vice-President of the Royal Society in 1804, and the lecture-going proprietors of the Royal Institution. Davy was to retain his hope of state-aided research institutions. (47)

A genuine desire to apply science was shown, since Banks, Greville and Hatchett all had excellent relations with mine owners, mineral surveyors and coal-viewers. Banks was always on good terms with the mineral surveyors, John Farey and William Smith, who both became critical of the Geological Society. Banks recognised that Farey 'founded his practise on an extensive and enlightened theory' to realise 'economic purposes'. Farey had combined his science with support for Malthus. (48) Smith's general notes presented geology as a means to strengthen society against materialism. He endorsed Banks's opinion of Farey, even though the latter had been given preference by the Board of Agriculture: 'In London Mr. Farey being put upon his own resources, his numerous published papers and his being sanctioned by the Board of Agriculture show the great progress he had made in the science before the Geological Society was established.' (49) Greville's Milford interests explain why he communicated the paper of the surveyor, Edward Martin, on the South Wales Coal Basin to the Royal Society. Although Farey sneered at this example of the patronage of a rival by 'the learned in mineralogy', Griffith declared his colliery survey of Ireland to be in agreement with Martin's map. (50) Smith felt the new profession of mineral surveyor was 'the legitimate offspring of Geology'. (51) Until such examples have received systematic study, the question of how geology came to be applied must remain open.

Some of the most exciting work on geology in the Royal Society was produced by Davy between 1805 and 1808. After extensive co-operation with fellow mineralogists, he discovered new properties of alkalis. He hoped these would lead 'to the solution of many problems in geology'. By revealing agents like natural electricity in the creation of minerals and in volcanic activity, 'this would hardly fail to enlighten our philosophic systems of the earth; and may possibly place new powers within our reach'. (52) He remained permanently interested in research relevant to mining, and he worked on detonating, combustion and the safety lamp. Davy was no amateur geologist, and Ian Inkster has justly reversed the view that there was only one professional geologist among the founders of the Geological Society, by pointing out that most had science-based occupations. (53) This shows a nascent professionalism in the Society, which, like mineral history, failed to develop.

Davy's geology suggests that theoretical concerns continued in a modified form. Even the sceptical Greenough admitted that 'a system is a good thing - it leads us to the observation of facts and it serves to tie them together'. (54) The ultimate aim of a general theory remained implicit in later particular studies. Kirwan, who was interested in primitive language, had conceived geology as the words, and mineralogy as the alphabet of the book of nature. Davy added that chemistry was the grammar. (55) Both Davy and Greenough were agreed that there should be a geological rather than a mineralogical society. A full account has been given by Martin Rudwick of Davy's quarrel with Greenough and how this led to a conflict over privileges between the Royal and Geological Societies. The dispute gave rise to the idea that geology was an independent science, and was not dependent on other sciences as Davy maintained. The significance of this becomes clearer when the Royal Society's constructive attitude to geology is recognised. Greville was not unreasonable in suggesting that the new society should be one of the Royal Society's assistant societies, with lower expenses to allow for a greater social breadth of membership. Greenough acknowledged that 'many good geological papers had appeared in the Phil. Trans.', but he criticised the Royal Society for its lack of a public reference library and collection, and because it did not sponsor research with non-members. (56) The Geological Society would also be a failure when judged by these criteria.

Fashion was not necessarily a positive factor in the rise of geology, since many of the new Geological Society members were less adept in the science. Most of the original nucleus of skilled researchers had tried to secure a judicious compromise with the Royal Society; Banks wrote to Greville of vociferous support in the new society. Yet, he felt, the dispute was being used to make the Geological Society a public name. He expected Babington and Bournon to resign along with Greville, Davy, Hatchett, Sir James Hall, and John Walker. (57) Greenough, with his immense private income, set the style of gentlemanly amateurism, which would increasingly prevail. Buckland was to marvel at the exclusiveness of the Society, and Jameson at its great wealth. Greenough was, in contrast to Davy, opposed to the application of science to mining, quoting Adam Smith as an authority. Whereas Davy saw organic matter regenerating the earth - so allowing for such economic applications of science as to soil improvement - Greenough saw, 'everywhere marks of decay, nowhere of renovation'. This was a pessimism matched by his belief in mankind's 'downright bestiality' necessitating a strong state church and an authoritarian government. (58) When Davy rejoined the Society in 1815, it was faction-ridden rather than a co-operative unity. Greenough, remembering the dispute with the Royal Society, was petrified. Warburton wrote to Marcet:

> The proposal of Davy at the G.S. has been met with such a spirit by certain persons, and they are so active in preparing for war, that D's friends must not be idle . . . Greenough & Macculloch after the meeting were loud in their murmurs, saying that the intention of the government was made manifest, viz: to reduce the G.S. to dependence on the R.S., and that this was the first act of a new system. In doing this they have manifestly put themselves in the wrong, and there is no inconsiderable want of candour & inaccuracy in such a statement. The Doctrine to be maintained is evident; that there were faults in the original quarrel; that Davy comes forward of his own accord without any suggestions from any part of the government, and that it must be regarded by every liberal person as an act of conciliation. Do not shew this to any Body; but preach the Doctrine with prudence. (59)

The undoing of mineral history

The attempt to institutionalise a structure for co-operative research only achieved limited success. This resulted in disenchantment with the world of the Geological Society. After the dispute with the Royal Society, committees were formed. The Committee for Chemical Analysis exhibited continuity with the idea of a mineral history. It was reponsible for mineral maps, bibliography, local investigations to ascertain the present art of mining in Great Britain and Ireland, and the general art of mineral surveying. (60)

The Swiss mineralogist, Berger, was commissioned to make surveys of particular localities by people such as Babington, W. Phillips and St Aubyn. But interest in chemistry and economic geology subsided. The residue of fieldwork came to be designated the 'English school'. Co-operation was done on an informal, individual basis. The Society had the minimum of professional staff and corporate enterprises. Its Transactions were established in preference to a mining journal. (61) A national survey and school of mines were both to be created in separate institutions.

The exclusiveness of the Society was criticised both from within and from outside. Mineral surveyors like Smith and Farey believed the Society exploited their experience without due acknowledgement. Fitton said there had been contacts with Smith only before the Society was founded. (62) But once again this twisted the record. Not only did Smith express interest in the Society's stratigraphy in his correspondence; the Society also visited his London collection :

> The Geological Society was then nominally in being - Mr. Greenough their President. I scrupled not to explain strata and the use of the Fossils so arranged as vouchers of the facts, not knowing but that the new body might be inclined to serve me: - in probability the maps were also opened and explained.

This account was written only after the publication of Greenough's map, which Smith resented as an infringement of the copyright of his own map of 1815;

> As a specimen of the liberality of public bodies (for such bodies are generally led by two or three Men) I may observe that they proceeded with other of my gratuitously instructed pupils as they did with Mr. Farey by making them honorary members and neglecting me - purposely it would seem to suit their sinister views. (63)

Farey attacked Greenough after his Ashdown section had been used as an abstract by Warburton. He also chided Greenough that the Society ought to adopt the high co-operation standards of chemistry. (64) Arthur Aikin and William Phillips made similar remarks, which echoed Davy's about the importance of mineral chemistry as the basis of geology.

Bournon had criticised the Society's Inquiries as dry and arid, lacking the aims of utility and teaching. (65) In 1820 MacCulloch criticised 'namby pamby cockleologists and formation men'. Doubts continued as to the competence of the Society. In 1816 Marcet observed 'fickleness' in the Society. In 1821 Horner recorded that 'I went to the Geological Society, which seems to me to have got into very feeble hands, and to want a great deal of the energy it had in former days'. In 1824 Dr Bostock feared the spirit of the Society would languish. (66) Of the Quaker mineralogists, Luke Howard later resigned, as did Pepys in 1829, and Allen, who was more concerned with the practicalities of his Lindfield Agricultural School in 1831. (67) That Aikin and Phillips remained active in the Society meant that a degree of interest in utility and in the Christian implications of the science were maintained. (Although overtly Anglican natural theology was unwelcome, the Society would reveal great prejudice against accepting evidence of man's antiquity.) The lack of co-ordinated research and the prevalence of factions were observed by Thomas Webster, who hoped several institutions would combine to endow a professorship of geology: 'In London there is as bad a set among the sçavans as can well be conceived, and the body of the sçavans are much too influenced by them . . . Most people are too indolent to probe things to the bottom, and thus the intriguers prevail.' This reinforces the view of a scientific society having a mass of thereby honoured but inactive members, whereas power lay in the hands of a clique. Webster feared that such organisations were 'never free from intrigue and that no situation can be held quietly but by keeping in (as it is called) with a powerful party, which requires an entire sacrifice of one's time and feelings, at least this is the impression produced by all my observations on the Royal Institution and the Geological Society, as well as what I hear of the London University etc.' (68) It was the peripheral mass of members who provide prime evidence of the influence of fashion and popularisation on the science. These were secondary factors in the establishment of geology, whereas social concern to apply science, when held by those who were themselves, or would employ, skilled researchers, was a primary stimulus to creating scientific advance.

Expectations of economic geology survived in the provinces and colonies. That the Society's _Inquiries_ were reissued by the government in Madras, suggests external demands on the Society contrasted with real capacity. Thomas Hardwicke hoped the Asiatic Society would co-operate with the Geological Society for economic ends. (69) The complex question of how science was applied must be distinguished from that of how consciousness of socio-economic circumstances produced a scientific response. It is with the latter problem that this paper is concerned, because the social context of mineral history is sufficient to suggest geology should be viewed as an aspect of the social forms and technology of industrialisation. A parallel to the London situation can be found in Ireland, where the government contributed first £5,500, and then £10,000, 'towards completing a cabinet of Irish minerals' and a laboratory for 'Experiments on Dyeing Materials and other Articles wherein Chymistry may assist the Arts'. (70) Davy also here 'turned the people's heads towards geology'. As at the Royal Institution, Bournon then failed to find employment, and an upstart faction challenged the scientific establishment by declaring itself to be the Mineralogical

Society.(71) Industrialisation thus divided rather than united the labourers of science. Smith felt that 'theoretical geology is the business of a particular class of people; geological practice is the business of an entirely different class'. (72) Two concepts of class directed science were in conflict: Banks's view of assistant societies fostering applied but apolitical science under the control of the Royal Society, or Greenough's preference for science as the privilege of leisured notables.

From an economic point of view, the method of establishing a detailed network of personal connections has been used to great effect by Musson and Robinson regarding the role of science in the Industrial Revolution. This, however, has been severely criticised by research drawing on sociological concepts of status and philosophical criteria for a distinct 'scientific' methodology. Yet Berman's suggestion that applied science was 'mass delusion' overstates the case. (73) An ideology of practical application, with science seen as a means of legitimating social status, offers much to explain the cohesion and disputes of the makers of geology. But this should not obscure the considerable efforts to apply the science. Historiographical controversy was symptomatic of the growth in activity and intellectual identity of geology. Previously the institutional structure of earth science was fragmentary and subordinate to other cultural and social concerns. Yet, with its diverse interpretations and centres of activity, early nineteenth-century geology failed to become a monolithic unity. To determine to what extent these sections were interrelated, the motivation of particular interests in the science, as well as the nature and diffusion of discoveries, need to be described. Later concepts of geology have obscured broader factors like the founding group of mineral historians' hopes for philanthropic and technical applications.

In the case of the mineral historians, apparently pure, descriptive research embodied a prescriptive programme of social action. Ideas of the necessary division of geology and mineralogy using criteria of the 1830s are misleading. Whewell was wrong to dismiss collectors as mere 'foster-mothers' of a geology 'born to rank and fortune'. His doctrine that geology and mineralogy were 'without any close natural relationship' was derived from philosophical idealism. (74) This opinion is contradicted by mineral history, as are Fitton's assertion that there was no English school of mineralogy and the idea that geology was more 'natural' than mineralogy and so more popular. Mineralogy was seen as an indispensable precondition of geology. Economic utility was an important incentive to research in these sciences, as a means of improving spiritual as well as material conditions, and also as a guarantee of truth in a divinely ordered world. These attitudes then changed in the Geological Society. Once the hopes of the founding group were not fully realised, the historiographic clash over what true geology ought to be arose. The Geological Society came to stand for 'the facts' - and indeed many were discovered by those who were its members. It was also an ideological stance to dismiss collectors as speculative theorisers and surveyors as crude empiricists. It was propaganda to bury the controversies of its immediate past and to conceal the demise of combined field, industrial and laboratory mineral history. There was a general sense of guilt in Horner's confession : 'If ever I am hanged for stealing, it will be in the cause of the Geological Society.' (75)

Though Smith realised that 'the application of a science to its practical uses is nearly as herculean as its establishment', without at least the intention of utility, geology would never have been established as it was. It is all too easy to pity the plumage of a science and forget the dying bird, the community of scientists, that produced it. Greville and his associates show the extent to which geology was a response to socio-economic needs. In contrast, the Geological Society failed in the field of social regeneration. The Society lacked a corporate spirit of economic utility dependent on co-ordinated research and teaching. It was therefore the fossil of a dying species of social ideals. Geologists cannot be seen as motivated purely by facts, as if they had hearts of stone. Nor, at the other extreme, is personal social status in itself an adequate explanation of their activities. The limits of both these interpretations were transcended by the mineral historians' ideals of man's social purpose on this earth.

Notes

I wish to thank the Royal Institution for allowing me to consult and quote from their archives.

1 H. B. Woodward, "The history of the Geological Society of London", London, 1907, p.52.

2 For general historiographical issues in the history of geology see R. S. Porter, 'Charles Lyell and the principles of the history of geology', Brit. J. Hist. Sci., 1976, ix, 91-103. Controversy over the specific points mentioned will be dealt with in the course of this essay.

3 Woodward, op.cit. (1), pp.6-10, 52. A. Geikie, 'The state of geology at the time of the foundation of the Geological Society', in W. W. Watts (ed.), "The Centenary of the Geological Society of London", London, 1909, pp.107-31 (110, 127).

4 M. J. S. Rudwick, 'Hutton and Werner compared: George Greenough's geological tour of Scotland in 1805', Brit. J. Hist. Sci., 1963, i, 117-35.

5 M. J. S. Rudwick, 'The foundation of the Geological Society of London: its scheme for co-operative research and its struggle for independence', Brit. J. Hist. Sci., 1963, i, 325-55.

6 R. S. Porter, "The making of geology: earth science in Britain, 1660-1815", Cambridge, 1977. Since completion of this paper, R. Laudan, 'Ideas and organisations in British geology: a case study in institutional history', Isis, 1977, lxviii, 527-38, has made a similar observation to that made here regarding the initial lack of success of the Society, although the reason for this is given as owing to the inadequacies of the Baconian inductive method. My view is that the Baconianism of a leisured amateur like Greenough had different methods and intentions than that of mineral historians interested in applied science.

7 Minutes of the Geological Society, i, section I, 'Of its objects'.

8 'Mémoire par le comte de Bournon', Library of the Geological Society MS., E Tracts 1.

9 Porter, op.cit. (6), p.132. R. S. Porter, 'The Industrial Revolution and the rise of the science of geology', in M. Teich & R. M. Young (eds.), "Changing perspectives in the history of science", London, 1973, pp.320-43 (338-9).

10 M. Berman, "Social change and scientific organization. The Royal Institution 1799-1844", London, 1978, p.71.

11 Journal of a tour to Ireland in 1806, Greenough Collection, University College, London. A. Aikin, "Journal of a tour through North Wales and part of Shropshire, with observations in mineralogy, and other branches of natural history", London, 1797. W. Phillips, "A selection of facts from the best authorities, arranged so as to form an outline of the geology of England and Wales", London, 1818, p.193, about his tour in 1800 'the inducements to which were the objects of mining and metallurgy'. P. G. Embrey, foreword to "Manual of the mineralogy of Great Britain & Ireland by Greg & Lettsom. 1858. A facsimile reprint with supplementary lists of British minerals", Broadstairs, 1977, p.viii.

12 E. Halévy, "England in 1815" (tr. by E. I. Watkin & D. A. Barker; intro. by R. B. McCallum), 2nd edn., London, 1949, pp.558-9.

13 E.g. the first version of this paper, discussed at Cambridge in 1977, contained an account of the British Mineralogical Society; this account is to appear in a revised form in: I. Inkster & J. B. Morrell (eds.), "Metropolis and Province. British science 1780-1850" (forthcoming). The scientific activities of Bournon and Greville will be assessed jointly by Peter Embrey (Dept. of Mineralogy, British Museum (Natural History)) and myself.

14 Greenough Papers, Cambridge University Library, Horner to Greenough, 26 June 1809.

15 E.g. Greville proposed Erasmus Darwin for the Royal Society, and they exchanged minerals and technical information. Watt to Black

3 March 1779, in E. Robinson & D. McKie (eds.), "Partners in science. Letters of James Watt to Joseph Black", London, 1970, p.56.

16 C. Greville, 'On the curundum stone from Asia', Phil.Trans, 1798, pp.403-48 (424). 'Advice for a tour to North Wales', (n.d.), Hamilton and Greville Papers, National Library of Wales. A. Raistrick (ed.), "The Hatchett diary. A tour through the counties of England and Scotland in 1796 visiting their mines and manufactures", Truro, 1967. British Museum Add. MSS. 42071, f.89.

17 R. Fenton, "A historical tour through Pembrokeshire", 2nd ed., Brecknock, 1903, p.104. A. Morrison, "The Hamilton and Nelson papers", 2 vols., 1893-4, Greville to Hamilton, 9 November 1792, nr. 214.

18 E. Laws, 'True history of Milford', Temple Bar, 1890, lxxxviii, 403-11 (409), Greville to Lady Cawdor, June 1808.

19 Greville papers, India Office Library, E 309. Greville's promotion of geology may be seen in the context of J. A. Schumpeter, 'The creative response in economic history', J. Econ. Hist., 1947, vii, 149-59 (152). Despite the limited use of science as judged by J.R. Harris, 'Skills, coal and British industry in the eighteenth century', History, 1976, lxi, 167-82, D. S. Landes, "The unbound prometheus", Cambridge, 1969, p.63, warns against 'the unlettered tinkerers of historical mythology'.

20 Porter, op.cit. (8), p.331.

21 'Societies for encouraging and promoting arts, manufactures, etc.', "Encyclopaedia Britannica", 3rd edn., vol. xvii, Edinburgh, 1797, pp.587-8. I. von Born & F. W. H. von Trebra (eds.), "Bergbaukunde", Leipzig, 1789, i, lists the members of the English section as follows: 'Director Herr John Hawkins zu London. Ausserordentliche Mitglieder: Herr Samual Vaughan der jüngere vorjetzt in Philadelphia in Nordamerica, Herr Peter Woulff vorjetzt in Paris, Herr Raspe in Cornwallis, Richard Kirwan Esq in Dublin, Herr Withering in Birmingham, Herr Tenant Chemist in Yorkshire, Herr Dr Hume zu Edinburg in Schottland, Herr Bolton zu Birmingham, Herr Watts zu Birmingham. Ehrenmitglieder: Herr Carl Greville zu London.' Hamilton was placed in the latter class in the Italian section. Further evidence as to the popularity of visiting mines is shown by the presence of other British travellers in Schemnitz in 1786. These were, besides Hawkins, Straton, Warren, Captain Wale, Roche, and Manners. Permission for their visits was granted by the Imperial administration, Vienna: Hofkammerarchiv, M. u. B., rote Nr. 2392. This was obtained through the British ambassador R. Murray Keith; British Museum Add. MSS. 35536, ff.45, 176, 225, 341, 358. Keith acted as Greville's agent to obtain minerals and also promoted the products of Wedgwood. Woulfe, Withering and Kirwan had been sent specimens of corundum and muriate of lead by Greville, as

stated in his Royal Institution proposal. Greville's activities supplement the discussion in M. Teich, 'Born's amalgamation process and the international metallurgic gathering at Skleno in 1786', Ann. Sci., 1975, xxxii, 305-40.

22 W. Hubatsch (ed.), "Frhr. vom Stein, Briefe und amtliche Schriften", 10 vols., Stuttgart, 1957-64, i, 261.

23 M. W. Flinn (ed.), "Swedenstierna's tour in Great Britain, 1802-3. The travel diary of an industrial spy", Newton Abbot, 1973, p.27.

24 Morrison, op.cit. (17), i, 145, Greville to Hamilton, 1790.

25 Royal Institution, Managers' Minutes, M3, p.295.

26 W. Babington, "A new system of mineralogy in the form of a catalogue", London, 1799. J. Carswell, "The prospector, being the life and times of Rudolf Erich Raspe (1737-1794)", London, 1950, p.163. Greville, op.cit. (16), pp.423-4. L. Namier & J. Brooke, "The House of Commons, 1754-1790", 3 vols., London, 1964, ii, 550-1 (about Greville),iii, 397-8 (about St Aubyn). For the latter see also, "Dictionary of national biography".

27 "Dictionary of national biography", (about Hume). Namier & Brooke, op.cit. (26), ii, 652 (about Hume). W. T. Brande, "A descriptive catalogue of the British specimens deposited in the geological collection of the Royal Institution ", London, 1816, introduction. British Museum Add. MSS. 42071, f.112, 'a chemical school' was deleted although intended.

28 Percy to Greville, 5 December 1782, Greville papers, loc. cit. (19), Packet B.

29 Hastings to Greville, Fort William, 15 March 1780, Morrison op.cit. (17), nr. 90. For examples of geology in the context of exploration, see B. Smith, 'European vision and the South Pacific, 1768-1850', Oxford, 1960.

30 Smith, 'Early history of geology', 1836, MS, Smith Collection, Dept. of Geology and Mineralogy, Oxford University.

31 Royal Institution, Box 14, 146.

32 Berman, op.cit. (10), p.89. Berman's account of the scheme contains further information (pp.88-92) but he does not mention Bournon or the Geological Society. For general discussion of the economics of the landed interest, see G. E. Mingay, "English landed society in the eighteenth century", London, 1963, chapters V-VIII.

33 Berman, op.cit. (10), pp.88, 95.

34 Rudwick, op.cit. (5), p.326. G. B. Greenhough, 'President's Anniversary Address for 1834', Proc. Geol. Soc., 1834, ii, 42-4 (42). Greenough papers, loc.cit. (14). Laird to Greenough, 3 July 1834; Laird's rejection of Greenough's account suggests that other formative influences may be shown on the Society, but also members' own sense of its history was unreliable. For Bournon : Halévy, op.cit. (12), p.564.

35 J. L. de Bournon, "Traité complet de la chaux carbonatée auquel on a joint une introduction à la minéralogie . . . ", London,

1808, p.x. The subscribers were: C. F. Greville and A. Hume (both were singled out for special thanks, p.xxxv), J. St Aubyn, Wm. Allen, Wm. Babington, R. Ferguson, G. B. Greenough, C. Hatchett, Luke Howard, R. Knight, R. Laird, Wm. and R. Phillips, J. Williams. Laird objected that Davy and Aikin were not among the subscribers, yet the former was associated with the Royal Institution, while the latter was a member of the British Mineralogical Society: Greenough papers, loc.cit. (34). Nearly 500 subscribers are listed in: "Subscription to the mineralogical collection and office of assay", Royal Institution Papers.

36 Banks to B. Thompson, (Count Rumford), 6 June 1804, in W. R. Dawson (ed.), "The Banks letters, a calendar of the manuscript correspondence of Sir Joseph Banks", London, 1958, p.816. Greville associated with the Pembrokeshire gentry to suppress disorder: Greville to Banks, November 1792, British Museum Add. MSS. 33979, f.187.

37 Banks to Thompson, April 1804, in Dawson, op.cit. (36), p.516.

38 Royal Institution, loc.cit. (25), M4, p.28 and passim.

39 Porter, op.cit. (6), p.206. D. M. Knight, 'The vital flame', Ambix, xxiii, 1976, 5-15 (10). For the general context of counter-revolutionary thought, see P.H. Beik, 'The French Revolution seen from the Right. Social theories in motion, 1789-1799', Trans. Amer. Phil. Soc., 1956, n.s.xlvi, 1-122. Bournon contradicts the idea that émigrés rejected Baconian empiricism; cf. F. Baldensperger, "Le mouvement des idées dans l'émigration française, 1789-1815", 2 vols., Paris, 1924, ii, 249. Greville also employed the émigré Jean-Louis Barrallier as civic and naval architect at Milford. For more general discussion of the impact of counter-revolution on science, see N. Garfinkle, 'Science and religion in England 1790-1800: The critical response to the work of Erasmus Darwin', J. Hist. Ideas., 1955, xvi, 376-88.

40 Greenough, op.cit. (34), p.42.

41 Berman, op.cit. (10), pp.2, 32, 58-61, 70-1, 88-92.

42 Minute Books of the Board of Agriculture, B 6., 1 May 1801, Institute of Agricultural History and Museum of English Rural Life, University of Reading. Banks proposed a work on agricultural botany to be undertaken by an agriculturalist, botanist, draughtsman and colourist, and surveyor; in February 1802 Banks deposited engravings of English noxious insects with the Board; on 2 June 1802 Banks acted as intermediary between the Royal Institution Professors and the Board; cf. also 15 February and 6 June 1803; 2 April and 21-2 May 1805.

43 "Communications to the Board of Agriculture, on subjects relative to the husbandry and internal improvement of the country", 7 vols., London, 1797-1813, iv, pp.xxxl-ii, 313. "The correspondence of . . . Sir J. Sinclair, with reminiscences of the most distinguished characters during the last fifty years", 2 vols., London, 1831, ii, 313.

44 Berman, op.cit. (10), p.49.
45 Minutes of the Board of Agriculture, loc.cit. (42), 22 April 1806.
46 Banks to Rev. M. Hitchins, 27 May 1800, Royal Society Letters and Papers 155. Cf. Greville to Hamilton, Morrison, op.cit. (17), nr. 270: 'the public will derive more benefit than myself from these works'.
47 D. M. Knight, 'Chemistry in palaeontology: the work of James Parkinson (1755-1824)', Ambix, 1974, xxi, 78-85 (83). "Dictionary of scientific biography", entry on Davy.
48 J. Banks, 'Effect of the Equisetan Palustris upon drains', Communications to the Board of Agriculture, op.cit. (42), ii, 349. Letter by Farey , Monthly Mag., 1804, xviii, 188-90, (10 September).
49 Smith, MS Autobiography, 1817-18, Smith Collection, loc.cit. (30).
50 Farey to Smith, 23 May 1806 and 14 February 1807, Smith Collection, loc.cit. (30), Griffith to Greenough, 4 October 1809, loc.cit. (14).
51 Smith, op.cit. (30).
52 Davy, 'The Bakerian Lecture on some chemical agencies of electricity', Phil. Trans., 1807, pp.1-56 (55).
53 I. Inkster, 'Science and society in the Metropolis: a preliminary examination of the social and institutional context of the Askesian Society of London, 1796-1807', Ann. Sci., 1977, xxxiv, 1-32 (20, 28, 31-2).
54 Diary of tour of Scotland, 1805, 14 October, Greenough Collection, loc.cit. (11).
55 R. Kirwan, 'Geological essays', London 1799, p.111. R. Kirwan, 'An essay on the primeval language', Dublin, 1805. Royal Institution, Early Papers, Box 26, Folder 43, 1814,? lecture notes, 'On geology'.
56 MS history of the Geological Society, Greenough Papers, loc.cit. (14).
57 Dawson, op.cit. (36), p.371. E. Edgeworth also threatened to resign, letter (n.d.) in Greenough Papers, loc.cit. (14).
58 Diary of tour of Scotland, 1805, 8 August and 10 September. Letters of Jameson to Greenough, November 1811 ('The Geological Society is so wealthy and its members have so much influence'), May 8 1816 ('You have ample funds greater members'); Political Journal, p.86, Greenough Papers, loc.cit. (11).
59 Griffith Collection, University College, London, typescripts of Marcet letters, nr. 12, 21 May 1815.
60 Greenough Papers, loc.cit. (14), Geological Society Committee Papers.
61 Rudwick, op.cit. (5), p.353.
62 W. H. Fitton, 'Notes on the history of English geology', Phil. Mag., 1833, 3rd series, ii, 37-57 (52).
63 J. Phillips, "Memoirs of William Smith", London, 1844, p.216. Smith, op.cit.(30). Letter to Rev. B. Richardson (20 March

1808), T. Walters (17 March 1808) and J. J. Smith mention the Geological Society's interest in strata. On 7 March 1808 Farey invited Hall and Greenough to Smith's. On 10 March Smith reinvited Hall, Lowry and Sowerby. On 15 March Smith wrote to Sowerby to ensure their studies on fossils would not clash. Both Farey and Smith had London residences and so could not be honorary members, as Rudwick has observed.

64 Greenough Papers, loc.cit. (12), Farey to Greenough, 16 September 1810, 29 May 1813.

65 A. Aikin, "A manual of mineralogy", London, 1814, p.vii. Phillips, op.cit. (10), p.9. Bournon, loc.cit. (7). Letter to Babington, pp.44-6.

66 Woodward, op.cit. (1), pp.60, 76. K. M. Lyell (ed.), "Memoir of Leonard Horner", 2 vols., London, 1890, i, 81, 85, 174, 192. I wish to thank Pietro Corsi, Oxford, for bringing this particular source to my attention, and more generally for his stressing of the importance of precise definition of changes in intellectual attitudes in response to particular historical developments.

67 Woodward, op.cit. (1), p.63. Howard had presented a paper on geology to the Askesian Society in 1802, and was a founder member of the Geological Society.

68 Webster to G. Cumberland, n.d., British Museum Add. MSS. 36512, f.276.

69 Geological Papers communicated to General C. H. Hardwicke FRS. British Museum Add. MSS.9894, f.29, T. Hardwicke to J. H. Harrington, 10 August, 1818, Calcutta re. 'the first Geological Society of Great Britain'; British Museum Add. MSS.9894, f.84, "Inquiries into geology for the purpose of obtaining information of this interesting science in such parts of the world as may afford additional intelligence, particularly the East Indies. Circulated by the Geological Society in Great Britain, Madras, reprinted from the Government Press, 1816".

70 T. S. Wheeler & J. R. Partington. "The life and work of William Higgins, chemist (1763-1825)", Oxford, 1960, pp. 17-18.

71 Henry Joy to Greenough, 24 November 1811, Greenough Papers, loc. cit. (14). On 20 March, Joy described the founder, Dr Ogilby: 'He is the founder and the Pres. and the every thing. He was ambitious of the honour of establishing something before Jameson should arrive and collected two or three together, and by putting names on paper and applications etc. got some v. gd. people to involve themselves under his banner.'

72 H. Wendt, "Before the deluge",(trans. R & C. Winston), London, 1968, p.87.

73 Berman, op.cit. (10), p.39. Cf. Porter, op.cit. (9); A. Thackray, 'Natural knowledge in cultural context; the Manchester model', Amer. Hist. Rev., 1974, lxxix, 672-709. A. E. Musson & E. Robinson, "Science and technology in the Industrial Revolution", Manchester, 1969, deals with 'mineralogy' rather than 'geology'.

74 W. Whewell, 'President's Anniversary Address for 1839', Proc. Geol.

Soc., 1838-42, iii, 65-6, held on 15 February 1839, about A. Hume and his association with the Babington group.

75 Horner to Greenough, 26 June 1809, Greenough Papers, loc.cit. (14).

NOTES ON CONTRIBUTORS

DAVID ELLISTON ALLEN is on the staff of the Social Science Research Council (but contributes here in his private capacity). He is the author of *The naturalist in Britain: a social history* (1976), and is currently engaged on a history of the Botanical Society of the British Isles.

PATRICK J. BOYLAN is Director of Museums and Art Galleries, Leicestershire, and has special interests in Quaternary geology and archaeology, and the nineteenth-century history of Quaternary studies. He is currently working on a major review of the scientific work of William Buckland.

W. H. BROCK is Reader in the History of Science and Director of the Victorian Studies Centre at the University of Leicester. He is Hon. Editor of *Ambix*, the Journal of the Society for the History of Alchemy and Chemistry, and has published *The atomic debates* (1967), *Studies in Physics* (1972), and *H. E. Armstrong and the teaching of science, 1880-1930* (1973).

JOHN HEDLEY BROOKE is Lecturer in the History of Science at the University of Lancaster. His primary research interests include the conceptual foundations of organic chemistry, and science and religious beliefs. He has published detailed studies in both these areas, including several texts for the Open University course 'Science and Belief from Copernicus to Darwin'.

G.N. CANTOR teaches the history and philosophy of science at the University of Leeds. His current research relates primarily to optics and theories of the ether in the eighteenth and nineteenth centuries.

R. GRANT, of the University of Cambridge, is completing a study of James Hutton's idea of history.

L. J. JORDANOVA is Research Officer, Wellcome Unit for the History of Medicine, University of Oxford. Previously, she was a Research Fellow at New Hall, Cambridge. Her principal research interests are in the history of the bio-medical and human sciences in the eighteenth and nineteenth centuries. She is at present working on a longer study of environmentalism and medicine in the period of early industrialisation.

MARCIA POINTON was educated at Manchester University and held the post of Research Fellow at the Barber Institute of Fine Arts, the University of Birmingham, until 1975 when she went to the University of Sussex as Lecturer in History of Art. Her book *Milton and English art* was published in 1970, and she has also written many articles for scholarly journals in the field of nineteenth-century British art.

Her monograph on the artist William Dyce will be published shortly and a critical study of the work of William Mulready is in preparation.

ROY PORTER, Churchill College, Cambridge, is university Assistant Lecturer in History. His publications in the history of geology include The making of geology: earth science in Britain 1660-1815 (1977). He is currently working on a social history of eighteenth-century England.

MARTIN J. S. RUDWICK is Professor of the History and Social Aspects of Science at the Free University of Amsterdam. He is the author of The meaning of fossils (1972) and of several articles on the earth sciences in the early nineteenth century. He is currently using this historical material to explore a variety of ways of relating the cognitive and social dimensions of natural science.

HUGH TORRENS is Lecturer in Geology at Keele University with special interests in Jurassic palaeontology and stratigraphy. An early interest in the stratigraphy of the Bath area, whence Bathonian rocks are named, led to research into how the science of geology had developed in that area.

PAUL JULIAN WEINDLING read history at Merton College, Oxford, before taking an MSc in the history and philosophy of science at University College, London. After a year as trainee librarian at the Warburg Institute, he studied in Munich for a PhD on Oscar Hertwig (1849-1922), the biologist and social thinker, and he is now at the Wellcome Unit for the History of Medicine, University of Oxford.

INDEX

(Prepared by L. J. Jordanova)